HEROES ◆ OF ◆
ANNAPOLIS

THE TRUE STORIES OF NINETEEN BOLD MEN
AND WOMEN OF THE U.S. NAVAL ACADEMY, FROM
THE MEXICAN WAR TO THE WAR ON TERROR

DAVID POYER

WITH AN INTRODUCTION BY
HON. J. WILLIAM MIDDENDORF II

NORTHAMPTON HOUSE PRESS

HEROES OF ANNAPOLIS Copyright 2019, 2022 by David Poyer
All rights reserved • Earlier versions of these pieces appeared in *Shipmate*, the alumni magazine of the US Naval Academy
Cover design by Naia Poyer
Academy Seal courtesy of Naval Academy Athletic Association
First Northampton House Press trade paper edition, 2019
Revised and expanded, 2022
ISBN 978-1-950668-08-3, 978-1-950668-13-7, 978-1-950668-14-4
Library of Congress Control Number: 2018900723
9 8 7 6 5 4

PREVIOUS BOOKS BY DAVID POYER

TALES OF THE MODERN NAVY

Arctic Sea   Deep War   The Threat
Hunter Killer   The Command
Onslaught   Black Storm
Tipping Point   China Sea
The Cruiser   Tomahawk
The Towers   The Passage
The Crisis   The Circle
The Weapon   The Gulf
Korea Strait   The Med

TILLER GALLOWAY

Down to a Sunless Sea   Louisiana Blue
Bahamas Blue   Hatteras Blue

THE CIVIL WAR AT SEA

That Anvil of Our Souls   A Country of Our Own
Fire on the Waters

HEMLOCK COUNTY

Thunder on the Mountain   As the Wolf Loves Winter
Winter in the Heart   The Dead of Winter

OTHER BOOKS

On Politics and War (with Arnold Punaro)   Stepfather Bank
The Whiteness of the Whale   The Return of Philo T. McGiffin
Happier than this Day and Time   Star Seed
Ghosting   The Shiloh Project
The Only Thing to Fear   White Continent

CONTENTS

Introduction by Hon. J. William Middendorf II.....................i
Foreword: The Mission of the Naval Academy......................1
1. Heroes Before Annapolis: The Midshipmen's Monument........7
2. Robley D. Evans, Class of 1863; Our Toughest Graduate.......19
3. Charles W. "Savez" Read, Class of 1860: From Anchor Man to Immortality..35
4. The Mid Who Moved the World: Frank Julian Sprague, Class of 1878 55
5. Philo N. McGiffin, Class of 1882: American Mandarin..........77
6. Richmond P. Hobson, Class of 1889: The Most Kissed Man in America...105
7. Wendell Neville, Class of 1890: "Retreat, Hell! We just got here!" 121
8. Merian C. Cooper, Class of 1915 (not graduated): The Man Who Captured Kong..135
9. Howard Gilmore, Class of 1926: "Take her Down."............157
10. Victor 'Brute' Krulak, Class of 1934: The Varieties of Truth .....
....177
11. Paul Shulman, Class of 1945: "If You Can Sink Them, Shoot"...
....203
12. Thomas Hudner, Class of 1947: He Could Be Counted On.......
....219
13. John Ripley, Class of 1962: "Hold and Die"..................239
14. Megan McClung, Douglas Zembiec, Erik Kristensen, Class of 1995: Heroes of the War on Terror........................261
Afterword: What is a Hero Today?..............................281
Endnotes...285
About Northampton House Press................................327

HEROES OF ANNAPOLIS

INTRODUCTION
BY J. WILLIAM MIDDENDORF II

A truism I stumbled on somewhere in my early reading told that what ultimately holds our life together is consciousness, and that we too often stick consciousness in our pocket. A military education teaches respect for the imperative of life, "You will." The epic personalities in the pages you're about to read illuminate a paradigm of shimmering consciousness in action from start to finish. These men and women were born wide awake and the Academy's "You will" added resonance.

American naval annals bulge with tales of extraordinary deeds by impossibly brave officers and ratings who sacrificed themselves for the nation we call home, for their comrades and missions afloat and ashore. They've demonstrated not only fearlessness in battle, where shot and shell dismember and kill, but also moral intrepidity, where a contrasting point of view vigorously defended can quickly extinguish a career. Better to fawn and acquiesce? Not for the men and women immortalized in these pages.

The U.S. Naval Academy's contribution to this pantheon of physical and spiritual grit is out of all proportion to its size. The subtext of David Poyer's splendid new book is that Annapolis incubates truly Olympian men and women. Picking out a dozen-plus heroes from a line of eligible candidates that reaches back to 1845 is daunting, but these selections make the best of a tough job. We can quibble about absent names, but there is no begrudging the men and women who made the cut.

Add Poyer's exceptional skill in telling a tale to this list of terrific histories—all of them true—and you have a compelling, eminently readable book. The writing pulls you in unceremoniously by the throat, amidst the worst of fears, amid the dead and the dying. Ordinary mortals can only stand in awe, turn the page and plunge in again. This is not a litany of plucky ex-midshipmen climbing the promotions ladder. No "E-Ring admirals" here, and not everyone served in the U.S. Navy. Example: "Savez" Read, Class of 1860, who skated "perilously near" the Academy's 200-demerit limit but went on to fight with consummate skill and valor at sea. Alas, employed in a losing cause—the Confederacy.

Old Gimpy Evans, Class of 1863, "toughest SOB" to ever graduate from the Academy. Buffalo hunter and Indian fighter. A man who stayed loyal to the Union when the rest of his family chose the South. Shot in the chest and legs, he fought to stay in the Navy but never walked straight again. A hero of the Spanish-American War, who was publicly scolded for his "coarseness and vulgarity."

Philo McGiffin, Class of 1882, a suicide who died in shockingly poor health (weighing 90 pounds), and who left a farewell note to a Miss Phelps "with apologies ... for the row—it is the way all guns have. Au revoir." He grew roses, played the flute, and fought as commander of a Chinese battleship in a war that changed history.

Richmond Hobson, Class of 1889, "A mid who abstained from alcohol, prayed each evening on his knees, and read from the Bible," and who led a daring raid into a Cuban harbor. Full captain at age twenty-nine. Congressman, defender of women's rights and energetic proponent of prohibition, fierce critic of the electoral college and recipient of "the kiss heard round the world."

There would be no King Kong or much besides, without the astonishing, improbably gutsy, adventurous and lucky Merian Cooper, summarily ejected from the Class of 1915. He was the pilot who brought the big ape down from the Empire State building in New York City. In World War I he fought the Kaiser in the Army Air Service. When his aircraft was seen plummeting to earth in a blazing ball of fire, the Army listed Cooper as probably killed. In fact, badly burned, he was taken prisoner.

In the Second World War, Cooper flew with the Flying Tigers against the Japanese, and served as an influential staff officer with "Wild Bill" Donovan and General "Vinegar Joe" Stilwell in the Pacific theater. In 1943, harking back to his failure at the Academy, Cooper told a friend, "I was not a good midshipman. But the Naval Academy did a lot for me. Perhaps I wasn't good for the Naval Academy, but it was good for me."

The legend of Marine Corps bravery at Belleau Wood, France, in World War I hinges on the indomitable spirit, humor and guts of Wendell Neville, Class of 1890. Neville was also the first ashore at the invasion of Veracruz in the war with Mexico. Neville epitomizes two core leadership truths: officers cannot define themselves as being "in charge." An officer "is responsible." That means accepting accountability.

Howard Gilmore, average guy. Receding hairline and chin, a quiet man. Some described him as meek. Down-to-earth, got along with everyone. Class of 1926. In World War II he became a U.S. Navy legend on the conning tower of his stricken submarine, ordering "Take her down," in the full knowledge it would cost his life, but save his command and crew. Gilmore was the first U.S. submariner to receive the Medal of Honor.

Two men I counted as colleagues and personal friends make the cut in *Heroes*: Victor "Brute" Krulak, Class of 1934, and Thomas Hudner, Class of 1947. Tom lived near me, a neighbor. Krulak displayed extraordinary physical bravery in combat, but in my mind he stands out most vividly as a fearless teller of truths, most famously to President Lyndon Johnson during the Vietnam War. Both men showed audacity in battle. Krulak carried his courage through very high rank in the bureaucratic arena where the average man often becomes cautious and timid. One of two men of Jewish faith in this collection, Krulak overcame anti-Semitism, ferocious Imperial Japanese Marines, a devious JCS-instigated attempt to reduce the Marine Corps to a cipher, Korean War challenges and Vietnam War blunderings to rise as the only senior general to confront LBJ. Passed over by President Johnson to lead the Marine Corps, Krulak retired. But things came right in the end. His son, "Chuck" Krulak, Class of 1964, became the 31st Commandant in 1995.

At the time *Heroes* was being written, Tom Hudner (Class of 1947) was the Academy's only living Medal of Honor recipient, a detail he would never be first to mention. Rather, Tom told stories on himself, as when he launched in his Crusader F-8E with the wings still folded. Shipmate and fellow aviator was Jesse Brown, the Navy's first African-American combat pilot. Hit by Chinese ground fire in Korea, Brown crash-landed but was trapped in the cockpit when the canopy slammed shut on impact. Hudner set his aircraft down next to Brown, who fell in and out of consciousness. For hours in the freezing cold Hudner attempted to pull his afflicted chum to safety, to no avail. Brown died of loss of blood and the cold. In 1972, *USS Jesse Brown* (DE 1089) was named in his honor.

The Class of 1945 produced the "quiet, introspective" (yearbook quote) Paul Shulman. Admiral at age 26. Founder and first Chief of Naval Operations of the Israeli Navy. In World War II his ship—*USS Hunt* (DD 674)—was struck by a kamikaze. Shulman recounted, "One thing you learn . . . When you're in the middle of combat you don't have time to be frightened. You can be frightened before or after." After the war, Shulman read about Britain's Royal Navy blockade of refugee ships taking Holocaust survivors to what was then Palestine. He resigned his commission and began a life that resembles the perils of Paul Newman in the 1960 film, *Exodus*.

The destruction of the bridge at Dong Ha, Republic of Vietnam, in 1972 has comparable naval historical prominence as the faces carved into Mount Rushmore. One man, John Ripley, Class of 1962, a devout Catholic and a Marine, stopped two enemy divisions—including two hundred Russian-made T-54 tanks—dead in their tracks and saved a country that "could not save itself." He did it with single-minded determination despite physical exhaustion, faint from loss of blood caused by razor wire, and intense enemy cannon and rifle fire. Dragging case after case of explosives to the bridge's trestles. "Je-sus Ma-ry get me there. Jesus-Mary-get-me-there. JesusMaryget me there...." With a great roar the bridge exploded sending a shock wave that knocked Ripley to the ground. He watched as A-1 Skyraiders took out stalled enemy columns.

Heroes ends with Megan McClung, Erik Kristensen and Douglas Zembiec, Class of 1995, who fell on the battlefield in the War on Terror. Megan, dead of a roadside bomb in Ramadi. Erik, a Navy SEAL, killed in a rescue attempt in Afghanistan, and Douglas, a Marine Force Reconnaissance platoon commander, by an enemy sniper in Baghdad. Selfless defenders of America, all. The Academy's "In Memoriam post 9/11" site maintained by the Alumni Association and Foundation lists fifty-nine men and women who were killed in operational losses, aircraft crashes, accidents and battle in America's longest war.

Heroes is an appropriately timely narrative with a truth far greater than a handful of noble names and deeds. It is a book about who we are as Americans; our values and aspirations. The mosaic of people chronicled in these pages constitute a fragment of our national saga. They are us, our fathers, our children our inheritance and legacy. Their example resonates in our past and lights our future with the promise that we have many more such as these. America's enemies, beware.

J. William Middendorf II
LT(jg) U.S. Naval Reserve, Engineering & Communications Officer, LCS-53, 1945–1946, Pacific
U.S. Ambassador to the Netherlands
U.S. Ambassador to the Organization of American States
U.S. Ambassador to the European Union
62nd Secretary of the Navy

FOREWORD:
THE MISSION OF THE NAVAL ACADEMY

The mission of the United States Naval Academy has never explicitly been to produce heroes. Officially, it is:

To develop Midshipmen morally, mentally and physically and to imbue them with the highest ideals of duty, honor and loyalty in order to graduate leaders who are dedicated to a career of naval service and have potential for future development in mind and character to assume the highest responsibilities of command, citizenship and government.

Note that nothing specific is said about courage, heroism, or self-sacrifice. Rather, it's about duty, loyalty, and command.

Yet it's often been written that courage is the first requirement of a military officer; that nothing else can be achieved without it. And embedded in the ideals of the profession, at least for those militaries of European derivation, are remnants of much older codes, derived from classical ideals, Judeo-Christian virtues, and medieval chivalry. This seems to be what the words "highest ideals" and "honor" refer to, in the Mission.

Now, certainly Annapolis has no monopoly on heroes—naval, marine, or otherwise. Nor are its graduates all stellar performers (or even, all stainless models of ethical behavior). But in all my researches, I've never found a single reference to a traitor, a coward, or a mutineer. Annapolis has graduated no Benedict Arnolds, Lord Haw-Haws, or Guy Burgesses. No Aldrich Ameses, John Walkers, or Robert Hanssens. No doubt there have been those who showed fear, who sought the rear rather than the front, or who did not step forward for the most dangerous assignment. But our records show no naval versions of J. Bruce Ismay, Francesco Schettino, or Eddie Slovik ... men who abandoned their responsibilities, deserted their posts, and let others die in their place.

This is an astonishing record, considering the thousands of graduates since 1845. It must mean that the Academy, opaque though its processes may seem while one is subject to their operations, is doing something right.

The Academy experience has changed in many ways since 1845. But one technique it's consistently employed is to hold up examples.

To this day, midshipmen are expected to memorize, and recite upon command, passages encapsulating the climactic moments of struggle in battle. "Don't give up the ship." "I have not yet begun to fight." "Don't cheer, boys, the poor devils are dying." "Damn the torpedoes, full speed ahead." They're constantly reminded that self-discipline and courage will be expected of them, both while in the Yard and after graduation. And examples and reminders are set before them on the grounds, in the form of statues, monuments, paintings, plaques, murals, captured flags, and most strikingly and solemnly, in a roll of honor in Memorial Hall, which commemorates graduates lost in action.

This idea of holding up heroic predecessors as models is no new conception. Plutarch, the Greek historian and biographer, employed the lives of heroes as teaching examples. "He believed that if young people study the lives of great heroes, and are taught to consider their virtues and flaws, they will naturally emulate them and use them as role models. He believed young people are driven by the passion for emulation, and he saw this passion at work in many of the heroes he wrote about: Alexander the Great emulated Achilles and wanted to imitate and exceed his feats; Julius Caesar emulated Alexander; and so on. Their emulation of previous heroes drove these figures on to great deeds."[1]

Of course, Greek ideas were different from ours. The classical definition of a hero was of a semidivine being, lower than the gods, but elevated above the common run of mankind.

Today, certainly, we no longer think of heros as semidivine, though we still use them as models, precedents, and at times, as warnings. And we still find the stories of their births, childhoods, struggles, and ultimate triumphs—or glorious defeats—endlessly fascinating.

But what, precisely, is a hero, today? How can we distinguish them from the merely brave, those who do their duty and no more?

It's time to define our terms.

The definition employed in this book is, loosely, of a warrior in battle who, at least once in his or her life, displayed individual courage and determination so far beyond the call of duty as to enlarge our understanding of what human beings are really capable of.

We'll explain a few other criteria for inclusion in a moment.

For the purposes of this book, we looked for men and women who met the following qualifications: engaged in battle in a personal way; displayed raw physical courage; been decorated, mentioned in dispatches, or otherwise publicly recognized; shown moral rectitude; and displayed what novelists Aleksandr Solzhenitsyn called civic courage, or John Ripley, moral courage.

Another lesson, perhaps, is the role of chance. As many, many decorated military people have remarked, they weren't that different from their shipmates or fellow troops. They weren't supermen. Just ordinary people, who just happened to be in the right place at exactly the right time . . . or the wrong time, from another point of view . . . but who then mustered the courage to do what they had to.

Perhaps we can learn two things from this. First, to recognize these rare moments when Fate demands our best, and understand our position at the hinge of events. And, second, to realize that how one acts at that crux may well define how others think of him, and how he or she thinks of him or herself, for the rest of their lives.

When the call for courage comes, each of us must be ready, whatever the arena we are tested in.

At some point, too, we must acknowledge all those whose names are not mentioned in this very compact book. This volume is very far from exhaustive! Hundreds, perhaps thousands of stories would have to be told for a complete record. Many obscure figures gave their lives as junior officers, and never had a chance to make their mark later in life. Others met their fates without witnesses, in a flaming gun turret, in the darkness of sinking, flooding ships, or alone in a trench. They fade from view unsung, but not unremembered.

But, from these thousands, I've selected a few for this particular book.

The point being: What can we learn from their lives?

From Philo McGiffin, perhaps, not to trust too much in the gratitude of a defeated and failing regime, even when one has served it well.

From Richmond P. Hobson, how quickly, and how completely, glory, and the most radiant public adulation, can evaporate.

From Gilmore and Ripley, how one's accomplishments can be appropriated, or reinterpreted, to forward the agendas of others.

From Robley Evans, how to acknowledge that at last, even the most heroic spirit may have to bend to the limitations of a failing body.

From "Brute" Krulak, how sometimes moral courage can be more quietly impressive, and even more difficult to muster, than the more obvious physical variety.

And from so many others, whom you'll meet, other lessons, deeper lessons, about life itself. And death, and how to face it and meet it, as we used to say, "like a man."

Before that, though, a note of thanks. This book grew out of articles originally published in *Shipmate,* the official magazine of the U.S. Naval Academy Alumni Association. *Shipmate*'s support, and particularly that of Kristen Pironis, my editor, made the writing possible. Scores of others also helped, far too many to note here. That doesn't mean they aren't appreciated. Their contributions are acknowledged in the endnotes at the back of this volume. I hope bringing these narratives to a wider public serves the common good. That is my personal goal in writing these pages.

And one more note: one of warning. While presenting these heroes as our examples, I've tried not to engage in hagiography, or to glorify war. Some of our subjects weren't good fathers, good sons, or good family men. Some were not really admirable, taken as a whole. Others didn't always act as nobly later in life as they once had.

Heroism is not a lifelong grace, and even heroes have their flaws, like any of us.

Still, for at least for one shining moment, each of the men and women in the following pages stood on the mountaintop. In that moment, they made the difficult choice. The hard choice. But as it turned out, in the end, the right one.

If we are to take heroes as our models ... let's see what we can learn from them.

 HEROES BEFORE ANNAPOLIS: THE MIDSHIPMEN'S MONUMENT

Their comrades raised a monument at their own expense ... to remember the friends lost in the last war that would be fought without Academy graduates in the ranks.

MARCH, 1847: VERA CRUZ, MEXICO

The heat was growing more intense by the hour, and the bites of sand fleas and other insects more maddening. The midshipmen and sailors, sweating in heavy blue wool uniforms, had slaved for a week preparing for today's bombardment. First they'd block-and-tackled the massive thirty-two-pounder naval guns out of the ships lying offshore. Lowered the seven-thousand-pound tubes into specially-constructed landing boats, then manhandled them out again at Collado Beach through heavy surf from a violent norther. They'd dragged those enormous unwieldy weights of iron, along with powder, shot, shells, and associated breeching and firing tackle, up the sandy beach and two miles over the dunes, while the wind blew stinging sand into their faces.

Now the storm had passed over, and here they were, before the port city of Vera Cruz and the Castle of San Juan de Ullóa. The ancient white tabby walls, fifteen feet high and studded with redans and forts, were holding the 12,000-man Army and Marine invasion force at bay. General Winfield Scott wanted to strike inland, assault Mexico City, and end the war before the yellow-fever season arrived. To do so he had been forced to call, reluctantly, on the heavy guns and skilled gunners of the Navy's Home Squadron.

Now, at the Naval Battery, twenty-one-year-old Passed Midshipman Thomas Branford Shubrick bent to peer along the barrel of his heavy gun. He'd taken the angular height of the fortification with a sextant, estimated the range, and calculated the proper elevation to hit the center of the wall seven hundred yards distant. Bullets whined overhead from the fort, now and then cracking into the ramparts young engineer Robert E. Lee had thrown up from rocks, logs, and the debris of smashed houses.

Shubrick tried to ignore the Mexican bullets. Had his uncles felt this nervous, when they'd fought in Tripoli, served aboard *Chesapeake, Constitution, Viper*, and dueled with HMS *Guerriere?* Or his father, who'd served with Decatur, and battled Malay pirates in Sumatra? He could not let the Navy, and the proud state of South Carolina, down now. Fortunately his crew was thoroughly drilled. Though the foe sheltered behind stone walls instead of oaken bulwarks, the young officer had every confidence he could batter them out of their redoubt.

But he couldn't *see.* He straightened, dragging an arm across itching, sweat-stinging eyes, wincing as a louse bit ferociously under the too-tight blouse. Fresh fusillades of bullets sang past, accompanied by the deeper, more menacing hummingbird-buzz of a cannonball. The Mexicans were concentrating their fire, furiously trying to suppress the Naval Battery.

"I can't see the target," Shubrick yelled. "Too much damn brush in the way."

From the gun beside him, Midshipman Allen McLane jumped to his feet, took two steps, and vaulted through the embrasure. He staggered as he landed, but recovered and began tearing away the shrubbery with gloved hands, first clearing Shubrick's sight line, then his own. Shubrick crouched involuntarily as another cannonball hummed over. More rifle balls cracked into the rocks around McLane, spattering them both with stinging fragments of lead and stone. "Get back in here, Allen," he yelled. "They've got our range."

"Just a minute . . . can you see now?"

"Yeah. Yeah! Get in here! I'm about to fire."

As McLane vaulted back through the embrasure, Shubrick straightened again, shading his eyes as he peered toward the distant walls. Then crouched once more, to make sure of his aim.

Shubrick never saw the round shot that killed him. McLane recoiled in shock as his friend and classmate's headless corpse reeled across the platform. Shubrick fell, jerking and twisting as his life blood spurted, soaking into the dusty soil of Mexico. [1] [2] [3] [4]

The monument stands at the crossroads of Chapel and Stribling Walk, at the confluence of the level red brick pathways linking Bancroft Hall with the various academic buildings and the Chapel. In other words, at the Academy's very heart.

A slim central obelisk is surrounded by stylized marble carronades and truncated stacks of marble cannonballs. Outboard of the sculptured stone, yet still part of the monument, point four bronze muzzle-loading smoothbore cannon. Of Spanish manufacture, they're mounted on green-painted castiron display stands. The inscription on the monument reads:

To passed Midshipmen
H. A. CLEMSON.
and
J. R. HYNSON.
lost with U.S. Brig. Somers
off Vera Cruz
Dec. 8th, 1846
This monument is Erected
by
passed and other Midshipmen
of the U.S. Navy
as a tribute of respect
1848

To Midshipmen
J. W. PILLSBURY.
and
T. B. SHUBRICK.
the former drowned off Vera Cruz
July 24th, 1846
the latter killed at the Naval Battery
near Vera Cruz
March 25th, 1847
while in charge of their duties
This monument is Erected
by
passed and other Midshipmen
as a tribute of respect
1848

Someone, somewhere, was determined to prevent these names being swallowed by oblivion.

Of the four, Shubrick was definitely the best connected. Not only was he from a prominent South Carolinian naval family, his uncle, Commodore (later rear admiral) William

Branford Shubrick, had commanded the closing operations of the war on the Pacific coast,[5] capturing Guaymas and Mazatlán, and going on to senior positions in the naval squirearchy.[6] Thomas may have been among those anointed from birth for eventual flag rank. But obviously this quartet were, when they died, neither admirals, noted statesmen, nor famous scientists. Nor, actually, did they ever graduate from or even set foot at the then-new academy at Annapolis![7]

Why then were their names and deeds judged worthy to commemorate here in marble and bronze? How, youthful as they were, could they be seen to embody, in some mystic way, the essence of the new Academy's mission?

Few human institutions spring to life fully formed, like Athena from the brow of Zeus. The Academy at Annapolis was amalgamated from a previous shore training site, the Naval Asylum at Philadelphia, combined with the much older system, inherited from the Royal Navy, of training prospective officers at sea, in active duty ships. Thus, when the inaugural class convened at the "Naval School" at old Fort Severn, it was something of a hodgepodge: students transferred from Philadelphia, those ordered in from previous Fleet training, and new accessions, with a five-year curricula that alternated duty at sea with instruction ashore.[8] Franklin Buchanan, the first Superintendent, had a big job ahead if he hoped to weld these disparate elements into a functioning, coherent institution.

His work wasn't made any simpler when the United States declared war on Mexico only one year later, in 1846, over the border with Texas. Buchanan immediately applied for sea duty, but was ordered to stay put at Annapolis. However, many senior midshipmen were ordered to the Fleet, while others applied for active duty.[9]

"Secretary Bancroft... advanced the *(graduation – Au)* examinations from November to July, 1846, and ordered the successful candidates to sea at once, which gave them all an opportunity for active service. In October, 1846, he assembled at the school about one-third of the 1841 date, who were graduated in July, 1847, and were sent to sea again, while the war was still in progress. The second group of 1841, which reported in October, 1847, had already seen active war service. All midshipmen of the 1840, 1841, and 1842 dates were given an opportunity to serve at sea during the war without interrupting their courses at Annapolis."[10]

"Ninety midshipmen, mostly oldsters from the school and fleet, saw service in the war."[11] Sources differ, but it seems clear neither Clemson, Hynson, Pillsbury, or Shubrick had ever actually attended the new shore school at Annapolis. But they were definitely well known to many of those mids who had—of which, more later.

The Mexican War proper began along the Rio Grande, and was initially fought primarily on land by the regular US Army. Meanwhile, "While the Army was fighting in Texas, the Navy's Home Squadron, under the command of Commodore David Conner sailed for Brazos Santiago, an inlet a few miles north of the Rio Grande. "[12] A close blockade was clamped down all along both coasts of Mexico, to add to the pressure, cut off munitions or other supplies, and make possible reconnaissance in support of future landing operations.[13]

The first mid to die in the war was Passed Midshipman Wingate Pillsbury, who drowned shortly after hostilities began, in July 1846. Assigned to *Mississippi*, a new side-wheel steam frigate on blockade duty, he was sent in charge of the ship's launch to chase a suspicious sail off Vera Cruz. When his launch was capsized by a violent wind, Pillsbury clung to the overturned hull. "With his crew he was able to keep his hold upon the overturned craft until he saw that one of his men who could not swim was nearly exhausted; and then, in attempting to give his place, which was more secure, to the sailor, he was swept away by a heavy sea."[14]

In early 1847, 12,000 Marines, Army regulars, and volunteers landed south of Vera Cruz under the command of General Winfield Scott. Scott's strategy echoed that of Hernando Cortez, pushing from Vera Cruz through the middle of Mexico toward the capital, Mexico City.[15] This would be the largest amphibious landing in history—at least until Gallipoli, in World War I.[16]

Unfortunately, the Army's siege-train had failed to arrive, including the heavy artillery essential to battering down masonry forts.[17] After the troops were ashore, and during a furious norther, Commodore David Conner and his newly-arrived relief, Commodore Matthew C. Perry, visited General Scott to discuss how the Navy could assist the siege.[18] Previously, Conner had offered to land a naval battery to assist the lighter Army artillery in breaching the walls.[19] [20] Scott now accepted this offer, and they agreed on a battery of six heavy guns complete with crews.[21]

"Loss of *USS Somers,*" *from* London Illustrated News

When the opportunity came to man this battery, the mids of *Mississippi* rolled dice for the chance. Among the winners was Shubrick. The heavier guns of the Naval Battery seem to have been the deciding factor in ending the siege. The heavy cannonballs shattered the brittle coral-based concrete of the walls, and the naval shells mowed down personnel. Army mortars and Congreve and Hale rocket bombardments added to the smoke, din, and terror in the city.[22] Soon a breach in the wall was opened, every enemy gun within range was silenced, and hundreds of civilians were wounded and killed in the town.[23] Scott prepared an infantry assault, but following a war council, the city surrendered. Its commander, Brigadier General Juan Esteban Morales, escaped in a small boat.[24] [25] Without the battery, Scott might have been stymied at the outset of the Mexico City campaign, and the war greatly lengthened;[26] though the enemy was already weakened by the "Polkos", army regiments who revolted after Vice-President Valentín Gómez Farías's expropriation of Church property to finance defense.[27] [28]

What of Clemson and Hynson? Both were passed midshipmen assigned to USS *Somers*. This small brig, of course, had an indirect connection to the establishment of the Academy.

For those unfamiliar with the "*Somers* Mutiny": in 1842, acting as an experimental schoolship, she was returning from carrying dispatches to Africa when Commander Alexander Slidell Mackenzie arrested, then swiftly condemned and hung, three men he accused of mutiny. One, Midshipman Philip Spencer, was the son of the Secretary of War.[29] The case excited deep divisions in public opinion. Following a court martial, at which Mackenzie was acquitted, many—including influential naval historian James Fenimore Cooper—considered his act not as the rigorous suppression of an incipient and dangerous rebellion, but rather closer to murder.[30] Some sources point to the affair as being one reason Congress began to consider the desirability of more rigorous professional training, conducted at a single facility ashore.[31]

Somers had been operating in the Gulf since the spring of 1846, on blockade and other duty, under the command of one Lieutenant Raphael Semmes, destined for notoriety in a later war as captain of CSS *Alabama*. On December 8 of '46, a sudden squall capsized *Somers* while she was chasing a blockade runner. Some 32 crewmembers drowned; others were apparently washed ashore and captured, and seven were rescued by a French warship there as a neutral observer.[32] Among the drowned were two passed midshipmen, Henry A. Clemson, from New Jersey, and John Ringgold Hynson, from Maryland.[33] Both had attended the Naval Asylum and passed their exams there.[34] Park Benjamin writes, "Passed Midshipmen Clemson and Hynson insisted upon their men taking refuge in the one available boat, but before further aid could reach them, their ship sunk. Clemson clung to a spar, which he deliberately abandoned when he saw that it was inadequate to support all who were clinging to it."[35]

It's worth noting that these four were not the only midshipmen to lose their lives in the Mexican War. Passed Midshipman William Thomas died of disease aboard ship off Vera Cruz. A Passed Midshipman McClanahan, assigned to *Cyane*, off California, may have been shot at the fort at St. Joseph's during the Bear Flag Revolt.[36] Another, R. Clay Rogers, was captured ashore while attempting to blow up a Mexican powder magazine, and nearly shot as a spy, but managed to escape.[37]

But all was not heroism, tragedy, or intrigue. Stories filtered back to the mids left behind about less fatal, even comic events. "There was Midshipman Young, for instance, who had wandered off with a dispatch to the dragoon commander before Medellin on an ancient cavalry charger which he had not the slightest idea how to manage. He arrived just when the charge sounded. The old war-horse promptly obeyed the bugle, and Young of course went with him, thus gaining the unsought honor of leading the dragoons and the cordial compliments of the colonel upon his remarkable gallantry."[38] There was also Foxhall Parker, who solved the sticky riddle of how to get a dismounted 32-pounder out of his landing boat by knocking a hole in the bottom and offloading it that way.[39]

The Mexican War seemed to prove the marked superiority of professionally educated officers over hastily-enrolled volunteers or part-time militia, especially in the fields of artillery, engineering, and reconnaissance.[40] [41] Congress allocated funds for the improvement of Fort Severn that August, and the U.S. Naval Academy was *officially* founded the next year.[42]

Now we return to the monument itself—properly termed the Midshipmen's Monument, not the "Mexican" Monument, as it's often called.

The 56 students in the Yard as the war ended raised $1000 among themselves to commemorate their fellows lost in the war, making it the first remembrance instituted by the midshipmen themselves.[43] (The much older Tripoli Monument was not government funded either, but had been raised by subscription among the officers who'd served in the Barbary Wars.[44]) No record survives of who first moved the resolution, or who donated the most money. But the fundraising does seem to have been a collective effort, and to have kept cost in mind—the obelisk, in funerary design, both symbolizes the connection between heaven and earth, and is cheaper to construct than other structures.[45] The monument was designed and sculpted in Philadelphia, of Pennsylvania marble, and installed in 1848, close to what was then the Academy's shoreline.[46] Philo McGiffin fired the Mexican cannons at midnight with saluting charges from the *Reina Mercedes*.[47] The monument was moved inland, to its present location, in 1909, but has otherwise been left as it was built, making it one of only three or four structures in the Yard dating from before the Beaux-Arts constructions of the Flagg era.

It is said that now and then, even today, a plebe may occasionally be seen at ground arms in front of the Midshipmen's Monument, holding not a rifle but a broom or swab. Could the implement symbolize a cannon rammer? Could this be a ghostly memory of Shubrick, as he manned his gun before his violent death at Vera Cruz? A collective tribute, welling up from what we might term the deep subconscious of the Brigade?

Regardless, the annals of Annapolis still reverberate with the deeds of those first heroes.

ROBLEY D. "FIGHTING BOB" EVANS, CLASS OF 1863: THE TOUGHEST GRADUATE

"If you wish to preserve the peace of the world, give us more battleships and fewer statesmen." —Robley D. Evans, 1908.

Who was the toughest, gutsiest USNA graduate ever? There are significant challengers for the position. But hard-nosed though others may have been, most pale beside a legend from the nineteenth century—Robley D. Evans, from the class of 1864.

Everyone from presidents on down called him "Fighting Bob." Buffalo hunter and Indian fighter before he ever got to the Academy, Evans commanded his first ship at seventeen and was wounded four times at Fort Fisher before killing his tormentor. When surgeons tried to amputate his legs, he swore to kill whoever tried. Invalided for incurable wounds, Evans came back to fight again at the Battle of Santiago, and dragged himself aboard the Great White Fleet on two canes. Evans was gruff, cranky, boastful, and so foul-mouthed he was chastised in the *New York Times* for his profanity. But those who knew him loved him, and he's our only grad ever to have a presidential pet named after him.

Evans wrote about his early years in his bestselling autobiography, *A Sailor's Log—Recollections of Forty Years of Naval Life.* He was born in the mountains of western Virginia, son of a doctor and legislator. He writes, "When I was six years old I was the happy possessor of a gun, a pony, and a negro boy. The first I learned to handle with considerable skill . . . the pony . . . seemed bent on breaking my neck; and the coloured lad, my constant companion, taught me, among other things, to smoke and chew tobacco." Evans recalls wagon trains, camp meetings, and long, hard winters.

His father died when Robley was ten. He moved to Washington to live with his uncle, a lawyer and newspaperman often about the Capital. As he entered his teens, Evans was about to run away to sea when a congressman from Utah, a friend of his uncle's, asked him if he would like an appointment to Annapolis.

The catch was that Evans had to establish residence in Utah. Not only that, in those days there were no plane, rail, or even road links to the newly acquired territory.

Evans took a train as far as St. Joseph's, Missouri, where he lost his money in a baggage mishap. Still, he managed to join a pioneer wagon train headed west. "As I was very young and small, I was assigned to assist the cook in preparing meals."[1] He also hunted buffalo, learning to ride up beside them on his mule and dispatch them with two or three revolver shots.

The second day out from Fort Laramie, though, the train was attacked by Indians. After driving off the whites, they took what they wanted and burned the wagons. "After standing the Pawnees off for some time and killing a good many of them, we made our way back to Fort Laramie, where we managed to secure one wagon and some pack animals, bought a fresh lot of supplies, and continued on our way."[2] Later they were attacked again, by Blackfeet. This time, Evans got an arrow through his ankle and partway through his mule.

At Green River, Bannock Indians rustled their horses, but a party of friendly Snakes (Shoshones) retrieved them.

Chief Washakie

At the celebratory powwow, the famous Chief Washakie treated young Evans to whiskey from a powder horn and nicknamed him "Little Breeches." Striking up a chumship with one of Washakie's sons, Evans was invited along on a ten-day hunt with the Shoshones. Evans "belonged to the chief's mess, so to speak," and had an introduction to Indian wrestling and hunting, and the nomadic life of the Plains. As they were returning, Evans wrote, Washakie offered him one of his daughters and a place in the tribe, but "Indian life had less charm for me the more I saw of it."[3]

The trek continued. They crossed alkali desert, and hunted elk and mountain sheep. Then, at last, looked down on Salt Lake City.

In 1859 the "Utah War" between the US Army and Mormon settlers had just ended. President Buchanan had issued a blanket pardon.[4] Evans stayed with a Mormon family, and had an interview with Brigham Young, whose flower garden he'd damaged on a runaway pony. When he'd satisfied the residence requirements, he took a seat on the overland coach east, in July, 1860.[5]

He reported to USS *Constitution* in Annapolis as part of a class of 127. The plebes were at first isolated from the rest of the Academy at the end of a long pier. They slept in hammocks. Their initial training concentrated on English, seamanship, and gunnery.

The mids took the Oath in the fall of 1860 even more seriously than usual, as if acknowledging not all would be able to keep it. "Each and every one appeared to feel himself standing at the crater of a volcano which might at any moment commence an eruption."[6] The resignations began even before South Carolina seceded in December 1860.⁷ Throughout the winter, more left to "go South."⁸ Separatist sentiment became overt, with some mids wearing blue secession cockades. Arguments raged aboard *Constitution* and in the classrooms. Evans reports that, following the example of Lt. Hunter Davidson, Class of 1847, "Conferences were frequent and serious, but never in one of them was a disloyal word uttered ... so long as we were inside the academy limits, or until our resignations were accepted, we were officers of the navy and would behave as such."⁹

Meanwhile, Evans got into several fist fights. Though he wasn't large, "his short body was all muscle and agility. He never hung back when a fight was pending."¹⁰ When he stood up for a smaller boy, a passing officer thought Evans had threatened use of a knife, and he was confined for mutiny. This was no empty charge so soon after the *Somers* hanging, but Evans got off with a restriction.¹¹

Evans, like most of the Southern cadets, faced a choice, but his mother tried to make it for him. He wrote, "As soon as war was an assured thing, my family demanded that I should resign, come South and fight for my state; but it did not seem to me that this course was imperative." His brother had joined the Confederates, but Evans decided to stay with the old flag. The decision was almost taken out of his hands, though. His mother wrote out his resignation and sent it to the Navy Department. It was accepted, and "... without previous warning I found myself out of the service, despite my determination to stay in." He went to Captain Rodgers, who sent a telegram to Washington explaining the situation. After a suspenseful twenty-four hours, Evans was notified he could still put "USN" after his name, but like many others, he was now estranged from his family.

After *Constitution* was towed to Newport, and the Academy reconstituted there,¹² the senior classes were ordered into service, leaving only Evans's class. Since they'd had little contact with the older midshipmen, though, there was a significant break in tradition. This, in Evans's recollection, allowed the previously unknown practice of hazing to take root. They trained hard in seamanship. Each Saturday the sloop *Marion* went to sea fully manned by the mids, who put it through every evolution, including grounding and ungrounding, under then-Lieutenant Commander Stephen B. Luce's demanding eye.[13]

The 1862 summer cruise took place under wartime conditions, aboard the sloop *John Adams*. The mids cruised south to Port Royal, then put in at Hampton Roads and Yorktown, where they rode cavalry horses along mined roads. Evans's leave also had its share of excitement when he shot a burglar in his uncle's coal cellar in Washington.[14] He also ran into his rebel brother in an oyster restaurant, and had to decide whether to report him to the provost (he did, but with enough of a delay to let him escape); and spent time with the Army of the Potomac. (He was not impressed.)

That winter Evans was caught stoning a jimmy legs and put in hack for two weeks. In those days, though, restriction consisted of being immured in *Constitution*'s "dark room"—literally a dark room, without light or candle. Upon emerging, "for several days I could not do anything, for the light hurt my eyes dreadfully."[15]

That June the upper half of his class was commissioned, leaving him behind, but he got his fill of action during his summer cruise. He served as executive officer of *Governor Buckingham*, then commanded the famous racing yacht *America* during the frantic search for the privateer *Florida*. He returned to Newport, to be immediately commissioned in October, 1863. Evans wrote, "Our educations were not complete, but we knew enough to look out for a ship and stop bullets, which were the important things."[16]

Reporting aboard USS *Powhatan*, he was assigned as a gunnery division officer along with several classmates.[17]

USS Powhatan

Laid down in 1852, the sidewheel steamer had been Commodore Perry's flagship on his mission to Japan.[18] For the next year, they pursued privateers and blockade runners in Caribbean waters, fighting yellow fever, rebel sympathizers, heat, and filth, varied with pitched battles ashore with crews of British ships. Late in 1864, though, the steamer was ordered home. Evans had seen plenty of action, but it was trivial compared to what came next.

By late 1864, Fort Fisher, at the mouth of the Cape Fear River, guarded the last entrepôt for Confederate blockade runners. The "Southern Gibraltar"[19] held two thousand men and fifty heavy guns. *Powhatan*, with Evans aboard, took part in the first assault in December. Despite the explosion of the "powder ship" *Georgiana*, the Confederates repulsed it with heavy casualties on the Union side.[20]

Admiral David Dixon Porter persuaded Ulysses S. Grant that a second attempt might succeed. This time, "A massive bombardment by Porter's fifty-seven warships covered the landing of Federal troops north of Fort Fisher on 13 January. As (Major General Alfred) Terry organized his units for the assault Navy ships, including five ironclads, kept up a nearly continuous fire for fifty hours, dismounting Confederate guns, blasting holes in the wooden palisades, damaging the earthworks, and making sure that the southern defenders received little rest."[21]

During the nineteenth and twentieth centuries, sailors regularly trained as landing parties. Accordingly, Porter called for volunteers to help storm the fort. Since Evans was the officer of the deck when the request came in, he put his name down first on the list.[22]

On January 14th Evans, commanding the commodore's barge, helped land the Army troops in the first wave.[23] The next day, a little after noon, two thousand sailors and Marines landed a mile and a half north of the fort. The "Naval Brigade's" mission would be to assault the northeast bastion.

"After the brigade moved up to a position 1200 yards from the fort, the flagship's steam whistle blared the signal to launch the attack. Sailors and Marines dashed forward led by their young officers. The Confederate commander of the fort, Col. William Lamb, personally led the majority of his garrison to defend the northeast bastion in the upcoming action. Porter's ships intensified their bombardment and covered the Naval Brigade until the men were within 600 yards of the defenses."[24]

Evans wrote, "At three o'clock the order to charge was given, and we started for our long run of twelve hundred yards over the loose sand. The fleet kept up a hot fire until we approached within about six hundred yards from the fort, and then ceased firing. The rebels seemed to understand our signals, and almost before the last gun was fired manned the parapet and opened on us with twenty-six hundred muskets. The army had not yet assaulted, so the entire garrison concentrated its fire on us."[25]

Almost immediately, he wrote, the marines broke as a unit, but most continued forward, mixed in with the sailors. "A withering fire cut down many of the attackers, and Evans instinctively pulled his hat down over his eyes unwilling to view the flashing blue line along the fort's parapet."[26] Under heavy fire, and taking terrific losses, the sailors would hit the deck every hundred yards or so, then rise and charge forward again. Eventually, those who reached earshot could hear the Confederates cursing them.

At this point, Colonel Lamb mounted the parapet, and Evans "considered him to be within range of revolver." On firing, he was himself hit with a bullet that sliced through his chest, spinning him around. He felt, however, "That was no place to stop," and kept running forward.

"As we approached the remains of the stockade I was aware that one particular sharpshooter was shooting at me, and when we were a hundred yards away he hit me in the left leg, about three inches below the knee. The force of the blow was so great that I landed on my face in the sand." Lamb had instructed his best riflemen to aim for the Union officers.[27]

A classmate helped him staunch the blood. "My left leg seemed asleep, but I was able to use it." Evans was soon up and limping forward again. His men cut the wires of unexploded "torpedoes" (mines) and soon he and seven others were within the "stockade"—apparently a preliminary entanglement set up as we would use barbed wire today—and getting ready to scale the parapet.

At that point, "my sharpshooter friend sent a bullet through my right knee, and I realized that my chance of going was settled." He collapsed, bleeding profusely, only to observe that the rest of the assaulting force was retreating, leaving him and a few others stranded just below the parapet, which was still filled with Confederates.

Everyone else who'd made it through the stockade had been shot down too, either badly wounded or killed. Evans lay in the open, trying to bandage himself with the silk handkerchiefs he'd brought along—they served as both battle dressings and tourniquets. "In the meantime, my sharpshooter friend, about thirty-five yards away, continued to shoot at me, at the same time addressing me in very forcible but uncomplimentary language." He hit Evans again, in the foot. Suddenly angry, Evans rolled over and shouted back at him.

Another Confederate was handing his tormentor yet another loaded musket when Evans fired once with his Colt Navy revolver. "My bullet went a little high, striking the poor chap in the throat and passing out the back of his neck."

This shot, which a modern expert calls "possibly the single most remarkable combat pistol shot in U.S. military history,"[28] removed the immediate threat, but Evans still lay in the kill zone, surrounded by hundreds of wounded and dead. A Marine from *Powhatan* made it through the stockade and dragged him to a shell-hole, but then was himself fatally shot.

At this point, Evans passed out. Meanwhile, taking advantage of Colonel Lamb's concentration on the Naval Brigade, the Army troops had attacked the fort at a different point.

When he woke again, the rising tide was threatening to drown him where he lay shot, bleeding, and half-buried in the sand. He noticed a Marine some yards away firing at the fort from behind a sandpile. He asked the man to help him get to cover, and was refused. Evans wrote, "I persuaded him with my revolver to change his mind." Dragged to defilade, yet still under fire, Evans watched the Army finish capturing the fort. "While the Sailors and Marines were struggling and dying before the northeast point of the defenses, Federal infantry had overwhelmed and breached the fortifications at the other end of the line."[29] Retrieved that night with the other wounded, carried a mile and a half to a makeshift aid station on the beach, he was nearly killed again by a Rebel shell. He was taken out to gunboat *Nereus* the next morning to join hundreds of wounded aboard the fleet.

Evans was severely wounded in the chest and both legs below the knees, "wounds which would have cost him his naval career at the age of 19, had he permitted amputation."[30]

But he didn't. *Nereus* debarked the wounded in Norfolk, and Evans overheard the surgeons at the naval hospital there discussing his case. The next morning, when the head surgeon explained what he had to do, Evans demurred. The surgeon repeated his plans, and Robley again refused. This time, "I pulled the gun from under my pillow; I told him that there were six loads in it, and that if he or any one else entered my door with anything that looked like a case of instruments I meant to begin shooting, and that he might rest perfectly sure that I would kill six before they cut my legs off."[31]

Instead they set him aside to die. Evans suffered through fever, erysipelas, bedsores, hospital vermin, and lingering infections. But he was determined to pull through.[32] Leaving hospital in June, he completed a long, painful convalescence at the home of an uncle in Philadelphia, "but he was never again to have free use of his legs. He finally walked, in a 'peculiar gait of his own,' which won him the nickname of 'Old Gimpy.'"[33]

As the postwar Navy faded from the national consciousness, Evans did shore duty at the Philadelphia Navy Yard and in the Ordnance Department in Washington. After a brief at-sea tour aboard *Piscataway*, he returned to Washington.[34]

Evans had first met his USNA classmate Henry Clay Taylor's sister Charlotte when she visited her brother during their plebe year. Robley and Charlotte were married in 1871. "Naturally, Evans wanted to spend his first years of married life on the beach, and he obtained an appointment to the Faculty of the Academy."[35] He taught seamanship, tactics, and naval construction, while Taylor taught mathematics. One of their students was Albert A. Michelson, USNA 1873, who did better in math than in Evans's classes. Ultimately, Evans found teaching not to his taste. Despite his handicaps, his career lay at sea rather than in the classroom.

In 1873 he was ordered to *Shenandoah* as navigator, and Charlotte accompanied him to meet that ship at Gibraltar.[36] His service in the Med led him to the conclusion that war with Spain was inevitable. *Shenandoah* was involved in the response to the *Virginius* incident, of gun-running to Cuba. Evans then succeeded William T. Sampson as executive officer of *Congress*, serving in the Canaries, Africa, and the Med again before returning to Philadelphia for the Centennial Exhibition.[37]

In 1877 Evans co-patented a signal light for communications at night,[38] then assumed his first command, of *Saratoga*, a sail training ship at the Academy.[39] Now he was training midshipmen the same way Luce had trained him years before. More Washington service followed. He served on the Naval Advisory Board, and fought through a resolution that all new U.S. ships should be built of steel. Appointed Lighthouse Inspector, he got himself fired when he forbade employees to pay kickbacks to the party that had appointed them.[40] This put him in bad odor with certain politicians, and he took a leave of absence to work for the B&O Railroad, building steel bridges. He was able to return to duty in 1885, under President Grover Cleveland, becoming Chief Inspector for Steel for the new cruisers being built.[41] After chairing the Lighthouse Board, he assumed command of *Yorktown* (PG-1)—fittingly, an all-steel, twin-screw gunboat, brand new and first of her class.[42]

It was aboard *Yorktown* that he gained his nickname. While Chile was angry with Washington over non-recognition of a new government, a mob set on the crew of Winfield Scott Schley's *Baltimore* in Valparaiso. Sailors were shot by police, probably while "properly drunk," as Evans put it, but added, "When in this condition they were more entitled to protection than if they had been sober."[43] Throughout this prolonged international imbroglio and war scare, Evans maintained his presence in Chilean waters.

"... one day the Chileans were practicing maneuvers and torpedo use in the harbor. In so doing, they came very close to the *Yorktown*, seemingly to intimidate the Americans. Evans protested. The president of Chile replied that the Chilean ships could travel wherever they desired in Chilean waters. At this, Evans stated that the *Yorktown* was the property of the United States government, and if the paint of the ship was so much as scratched, he would sink the offending torpedo boat."[44] By the end of the standoff the papers at home were calling him "Fighting Bob," and the name stuck. Evans, though, thought he'd exercised praiseworthy self control. "In the discharge of my duty I gave the Chileans a fine chance to fight if they wanted to, and the odds were enough in their favor—nine ships to one. But they backed water every time."[45]

Evans now alternated between Washington and major sea commands. In 1889, as equipment officer at the Washington Navy yard, he co-patented a machine for cutting and bending steel links for anchor chains.[46] He commanded *New York*, our first armored cruiser, aboard which he attended the opening of the Kiel Canal; and *Indiana*, one of the three steel battleships laid down in 1890. He became friends with Grover Cleveland, taking the president and cabinet members for inspection trips on lighthouse steamers.[47] He also worked closely with the assistant secretary of the Navy, Theodore Roosevelt, and was instrumental in the Personnel Board's merging of the engineering and line officer communities in 1896.[48] As the Spanish-American war approached, he took command of *Iowa*, the Navy's newest and largest battleship.[49]

On May 12, 1898, *Iowa* and other ships of the squadron bombarded the Spanish batteries at San Juan. On July 3, she was blockading Santiago when Admiral Pasqual Cervera's fleet sortied from that harbor. Evans fired on the flagship, *Infanta Maria Teresa*, then on *Vizcaya*. He then took on *Cristobal Colon* and *Almirate Oquendo*. Evans wrote later that the latter ship "pluckily held on her course and fairly smothered us with a shower of shells and machine gun [fire]." After the squadron disposed of two Spanish torpedo boat destroyers, it became a running engagement along the coast. *Oquendo* and *Maria Teresa* were set ablaze and sank. Evans pursued *Vizcaya* until she struck her colors and ran herself aground.[50]

"With other ships of the fleet involved in the pursuit of the escaping *Cristobal Colon*, Evans chose to go to the aid of the crew of the *Vizcaya*. The Spanish crewmen, while trying to escape the burning vessel and climb onto the beach, were being attacked by the Cubans. Evans was incensed by this attack on defenseless men who had fought to the best of their ability. Lowering boats, a landing party was sent ashore to defend the Spaniards against the Cubans. An officer was sent to find the Cuban commander and inform him that "unless they ceased their infamous work," Evans would turn the immense guns of the *Iowa* on the Cubans themselves." [51] *Iowa* suffered no losses in the action.

Evans won public acclaim for his aggressiveness at Santiago, but also managed to get himself publicly chastised for his "coarseness and vulgarity," profanity, and boastfulness when he wrote an ill-advised letter to a Pennsylvania newspaper.[52,53]

After the war, Evans served as Chief of the Board of Inspection. He also served on the Radio Board that put wireless to use in the Fleet.[54] When his old friend Theodore Roosevelt became president in 1901, moving into the White House with a guinea pig named "Fighting Bob,"[55] and Evans was made rear admiral in the same year, it was obvious his rise was not over. Evans and Roosevelt developed a relationship rather like the nearly contemporary one of Admiral "Jacky" Fisher and Winston Churchill, on the other side of the Atlantic. Roosevelt sent him on special missions to Samoa and Hawaii, and asked him to escort Prince Henry and Tirpitz in their visit to the United States. (Evans posed a possibly ill-advised question to Tirpitz, asking why Germany wasn't developing submarines.[56]) Roosevelt then named him to command of the Asiatic Fleet.

In August 1907, now in command of the Atlantic Fleet, Evans was asked to Oyster Bay to a high-level conference.[57] This time the President wanted him to command an expedition that still lives in history: the cruise of the Great White Fleet.

Roosevelt was "pivoting" the US fleet into the Pacific to counter a rising empire irritated by California's discrimination against ethnic Japanese.[58] He discussed this again with an increasingly crippled Evans at Sagamore Hill. The two attended the Army-Navy game together in Philadelphia that November.[59] (Navy won, 6-0.)[60]

Evans ca. 1903

The Great White Fleet sailed from Hampton Roads that December. Evans's son Taylor was also with the fleet, aboard *Louisiana* as a lieutenant. "The armada was under the command of the picturesque 'Fighting Bob' Evans, who had assured his listeners at a farewell banquet given by the Lotus Club of New York that they would not "be disappointed in the fleet, whether it proves a feast, a frolic, or a fight.'"[61] Actually, Roosevelt feared the Japanese might attack the fleet en route, or lay a trap for it; he instructed Evans to be alert for mines during his stop in Rio de Janeiro and during the transit of the Straits of Magellan.[62]

"Fighting Bob" Evans

Evans was often confined to his bunk during the voyage,[63] but when a drunken Marine knocked down two Chilean officers at a party in Punta Arenas he cut loose with a flood of profanity on the unfortunate commanding officer. The Chileans decided to hush up the incident,[64] and Evans left it out of his memoirs;[65] it came out only after his death.

Even in pain, hardly able to stand, Evans took the fleet around the tip of South America, then the length of two continents again to California. There, however, he was forced to leave for surgical treatment ashore.[66] He had to ask Roosevelt to relieve him of his command for the rest of the voyage.[67] At a banquet in his honor, he was carried down from his hotel room for a short speech, during which he uttered the words used as an epigraph to this chapter.[68]

Returning from San Francisco, Evans served on the General Board.[69] He was transferred to the retired list in 1908, after forty-eight years of service, and died in 1912 at his home in Washington. "As the flag on every ship and station in the Navy was half-masted, there passed from cabin to forecastle the sad word that Old Gimpy was no more."[70] Evans is buried in Arlington Cemetery.[71] His tombstone reads: *The path to duty was the way to glory*.

Among the American public at large, Evans is hardly remembered today. Perhaps if he'd borne a less commonplace name we would recall him better. He may, though, live on in our pop culture through another meme. A short, gruff, profane sea dog, with a gimpy, swaying walk and muscular forearms; ever ready for a fight, and always victorious in the end—strike any bells?[72]

USS *Evans* (DD-78) was commissioned in 1918 and transferred to Britain in 1940. In 1942, Evans's daughter Charlotte christened the Fletcher-class destroyer *Robley D. Evans*, DD-552.[73] *Evans* received the Presidential Unit Citation "For extraordinary heroism in action as support destroyer on radar picket station number 15 during an attack by approximately 100 enemy Japanese planes, 40 miles northwest of the Okinawa Transport Area, May 11, 1945."[74]

Her crew always called her the "Fighting Bob," inspired, no doubt, by the toughest SOB ever to graduate from USNA.

3
FROM ANCHOR MAN TO IMMORTALITY: CHARLES W. "SAVEZ" READ, CLASS OF 1860

"America never produced a navy officer more worthy of a place in history." —Admiral George Dewey

Born in fever-ridden Yazoo County, Mississippi, in 1839, Charles William Read grew up far from the sea. His father died in 1849, en route to the California goldfields. His mother sold the farm and moved to Jackson, the state capital.

To help out, Charles worked as a printer's devil after school, like the young Mark Twain. It was hard work for an undersized boy—he never got above about five feet five—setting type, inking plates, and street-hawking *The Mississippian*. He wore his reddish-brown, unruly hair long, and liked to dress up and act in amateur theatricals. It's hard to say now what attracted him to the Navy, but he did apparently play a sailor in one dramatic role. In 1856 he wrote to his congressman and asked for an appointment to the Academy.[1]

He arrived there aged seventeen, but immediately ran into a problem. Read "...was not very scholastically inclined. He seldom opened a textbook, yet he managed to pass his courses but with only the barest of minimum grades."[2] He

Savez Read as a midshipman

admitted to learning only one word of French: *savez*, meaning "to understand." Either proudly or derisively, he tacked it onto the end of almost every sentence. This became his class nickname: "Savez" or alternately "Savvy" Read. Still short, still thin, and with hair down to his shoulders, he looked a bit like a young Bradley Cooper (*American Hustle, American Sniper*). Classmates said his manner was "respectful and gentle,"[3] but Read was no pushover. He received demerits for not sweeping his room, for throwing snowballs, cursing, and fighting, skating perilously near the 200-demerit limit.[4]

Summer cruises to the Azores, Plymouth, Brest, Madeira and Cadiz provided Savez's first experiences at sea.[5] Unfortunately, a lack of interest in academics persisted throughout his time by the Severn. In 1858, he was 26th out of 33.[6] In 1859, 25th out of 26.[7] And his final exams firmly seated him at the back of the pack. Out of a graduating class of 25 in June 1860, he came in 25th of 25[8] ... the proverbial Anchor Man.

During Read's first class year, the nation seemed to be coming apart around him. Though the mids tried to avoid political discussions, "Each and every one appeared to feel himself standing at the crater of a volcano which might at any moment commence an eruption."[9]

Shortly after his graduation, Mississippi became the second state to officially secede, on January 7, 1861. Read later wrote,

When I received intelligence that my native State, Mississippi, had by the sovereign will of her people, severed her connection with the American Union, I was serving as a midshipman on board the United States steam frigate "Powhatan," then stationed at Vera Cruz, Mexico. I immediately tendered my resignation . . .On the 13th of March we arrived and anchored off the Battery, in the harbor of New York. The following day I started for the South, and was soon in Montgomery, the capital of the Confederate States. I called on Mr. Mallory, the Secretary of the Navy, who received me kindly, and informed me that no doubt my services would soon be needed by the Government. I also called on Mr. Davis, with whom I was acquainted. He asked me many questions about the Naval Academy, and the naval service, and seemed anxious to know how the officers of the navy from the South regarded the secession of the States. He said he hoped there would be no war, but if coercion was attempted, that the army of the South would be the place for a young man with a military education.[10]

Read then went home to Mississippi, and found time during this brief idleness to fall in love with a local girl, Rozinda (Rosa) Hall. Read's family weren't slaveholders, and he was never, throughout his life, recorded as giving voice to racist views. But when word came that Lincoln had called out the army against the seceding states, "The youth of the South sprung to arms ... and everywhere the fife and drum were heard." Read almost followed Davis's advice and volunteered for the army, "but I remembered that the South had but few sailors and would need them all on the water."[11] On May 1, he reported to Captain Lawrence Rousseau CSN, in New Orleans, for duty on *McRae*.

Marques de la Habana had actually been an English-built ship purchased by Mexican revolutionaries. She was captured as a pirate by USS *Saratoga* in March 1860 and sold as a prize in New Orleans.¹² Bark-rigged, but with a steam engine, she was bought by the CSN, named after a member of the Confederate Committee on Naval Affairs,¹³ and commissioned in March of 1861, under command of Lt. Thomas B. Huger, CSN. *McRae* displaced about six hundred tons, and was being mounted with six thirty-two pounders, one nine-inch Dahlgren gun, and a twenty-four pounder brass rifle.¹⁴ Outclassed and outnumbered by the Union fleet, the Confederacy planned to reprise the strategy of cruiser raids on merchant shipping that had proved so expensive to the British in the War of 1812.¹⁵

But a lot had to be done, and *McRae*'s sailing was delayed until the Union blockade of New Orleans changed her mission. As part of Flag Officer G. N. Rollins's improvised force, her new assignment was the defense of the entrances to the Mississippi, which also meant protecting blockade runners as they slipped in and out of the passes.¹⁶ Huger assigned Read to head up the sailing department.¹⁷ But getting ready for sea took so long, with so many complications, that Read also volunteered for an expedition to Ship Island. There he and a lieutenant

CSS McRae, coaling at Baton Rouge, 1861

emplaced several guns and exchanged fire with a Union gunboat, finally driving her off.[18,19]

McRae was at last ready, and tried for the open sea. But her engines failed, and soon "All idea of running the blockade and going to sea as a cruiser had been abandoned ... for with her limited coal capacity, and her want of speed owing to the small power and uncertain humor of her gear engines, it is doubtful if she would have lasted a month in that business."[20] She was eventually sent up the Mississippi to aid in the defense of Island Number 10. We catch a vivid glimpse of Read when, in the "wretched management and infamous scenes that attended the evacuation,"[21] demoralized Confederate troops tried to storm aboard as they fled New Madrid:

> They found a sentry at the gangway who ordered them to halt. They raged and swore and openly threatened to rush the sentry, but at that moment the gentle "Savez" Read appeared on the scene and told the men that if they came on board it would have to be in an orderly manner as soldiers, and not as a mob. At this the men commenced to threaten him, but he only asked them where their officers were, and was told that they did not care a rap where they were, but that they were coming aboard. By this time Read had gone ashore and was standing amongst them. He quietly asked them to be silent for a moment, and then inquired who was their head man. A big fellow, with much profanity said he "had as much to say as any other man." Instantly Read's sabre flashed out of its scabbard and came down on the heart of the mutineer, felling him to the ground, as in a thunderous voice the usually mild "Savez" roared, "Fall in!"—and the mob ranged themselves in line like so many lambs and were marched quietly across the gangplank and on to the ship.[22]

McRae fought in the battle of the Head of Passes in the fall of 1861,[23] then again in the spring of 1862, resisting David Farragut's advance upriver to capture New Orleans.

In this second battle, as Farragut forced the passage to the city, *McRae* was hammered hard by the heavy guns of the superior Union fleet. She was aground and on fire when her captain fell, mortally wounded. "Captain Huger was struck in the groin by a grapeshot and afterwards his temple was laid open by a canister bullet. When taken below he pleaded with Mr. Read, saying, "Mr. Read, don't surrender my little ship. I have always promised myself that I would fight her until she was under the water!" And right gallantly did "Savez" Read keep his word to his stricken captain, for when day broke the *McRae* was the only thing afloat with the Confederate flag flying."[24] Taking over command, Read extracted *McRae* from the mud and put out the fires. Nearly alone, she pursued the Federal fleet upriver, until, hit savagely once more, she lost steering and ran aground again. (This was the action for which Read was mentioned in a joint resolution of the Confederate Congress, thanking him for his part in the battle.[25]) After that lost engagement, he took the wounded to New Orleans under a flag of truce, then scuttled the damaged *McRae* under Farragut's nose.[26]

The campaigns in the West, though not as famous as the bloody Eastern battles, were strategically decisive. Union control of the Mississippi would cut three states off the Confederacy and allow penetration of the interior, "defeating hostile armies and eating out the vitals of the country."[27] This the South had to resist, and Read, finally promoted to lieutenant in October 1862,[28] was in the thick of the fighting, first on *McRae*, then, as the Confederates lost fort after fort and retreated, aboard the ironclad ram *Arkansas*. Built at Yazoo City, this casemate ironclad had such poor engines that her most famous engagement consisted of a drift down the Mississippi.[29] After carrying out a daring reconnaissance downriver on horseback, to find where the Union fleets and batteries were located and to get orders from General Van Dorn in Vicksburg,[30] Read directed *Arkansas*' stern battery[31] during her epic cruise downriver through the two Yankee fleets then bombarding Vicksburg.[32] He fought aboard her until her eventual loss below that city.[33]

After a period of working on harbor defenses, then a break to recover from "swamp fever," Read achieved what he'd hoped for from the beginning. John Maffitt, commander of the famous commerce raider Florida, requested him by name.[34] "Being perfectly delighted with the prospect of getting to sea,"[35] "Savez" reported to Mobile that November,[36] to begin the most illustrious and action-packed phase of his career. Maffitt was pleased too; he logged, "November 4—Lieutenant C. W. Read, the last lieutenant I personally applied for, joined; this officer acquired reputation for gunnery, coolness, and determination at the Battle of New Orleans. When his commander, T. B. Huger, was fatally wounded he continued to gallantly fight the McRae until she was riddled and unfit for service."[37]

CSS Arkansas *by R.G. Skerrett*

In contrast to the scratch vessels Read had served on to date, *Florida* had been designed and built in Liverpool as a pure commerce raider.[38] Fast, heavily armed, and dangerous, she broke out through the Federal blockade at night during a storm in January, 1863, winning an epic daylong chase against USS *Cuyler*.[39]

"Operating in the Atlantic and West Indies over the next eight months, *Florida* captured twenty-two prizes, striking terror in the United States' merchant marine and frustrating the U.S. Navy's efforts to catch her."[40] Read, however, soon parted company with Maffitt. After *Florida* captured a small brig off the coast of Brazil, Read proposed that he should skipper the little *Clarence* into Hampton Roads and cut out or burn Federal shipping there. He left in May, 1863. *Florida* was eventually captured at Bahia by USS *Wachusett*.[41] But young Read, now a commander in his own right at twenty-two, was about to sail little *Clarence* into legend.[42]

Slowed by a foul bottom, Read's cockleshell crept eastward. Meanwhile, he cut fake gunports in her sides, and augmented the single twelve-pounder brass howitzer[43], six rifles, thirteen revolvers and ten pistols[44] Maffitt had given him with "Quaker guns" made out of wood.

His first capture was a schooner, *Whistling Wind*, laden with coal for Farragut. Read burned her off Hatteras. On June 9 he burned another brig southbound with military stores before being intercepted and boarded by one of his classmates from a Federal warship. The man didn't recognize Read, who'd grown whiskers since the Academy, and Read pretended to be the captain of one of the ships he'd captured.

After reading captured newspapers and interrogating prisoners, Read decided Hampton Roads was too well defended for him to attack as originally planned.[45] Also, he needed a bigger, faster ship than *Clarence*. Fortunately a large bark-rigged merchantman of three hundred tons,[46]

Tacony, sailed up just then. He hoisted a distress signal to make it heave to, then boarded and captured it.

During the nineteenth century and into the early twentieth, international usage—"cruiser rules"—permitted navies and even privateers to capture enemy merchant ships, and either destroy them or send them in to a friendly port for adjudication and later sale as prizes of war. But the lives of crews and passengers had to be safeguarded.[47] Capturing two more ships while transferring his crew and fake guns to *Tacony*, Read sent off his accumulated prisoners in one and burned the other.

Read's bold depredations just off the main Union base at Norfolk triggered a panic in insurance markets. The Secretary of the Navy, Gideon Welles, responded to frantic shipowners. Dozens of Union men-of-war sortied after the "pirate," from Boston, New York, Philadelphia, and Hampton Roads.[48] Things looked dark for Read, who after all still had only one functioning small cannon.

Sailing north, he took another prize off Delaware, then angled toward the transatlantic routes out of New York. He evaded Union gunboats in a heavy storm. Today we might say Read was stealth-shielded: His ship didn't look like a cruiser. Patrolling warships took him for a merchant. He doctored a log and instructed his crew to imitate sailors on one of the ships he'd captured, so that even a boarding and papers inspection wouldn't give them away.

By late June Read was off Nantucket, and captured a Black Ball sailing packet. Her captain offered to burn her, but finding 750 Irish aboard, Read knew he could not dispose of so many prisoners. He bonded the ship for $40,000 and let her go. He next captured and burned a fishing schooner, then a clipper, *Byzantium*, with a cargo of coal. Then a bark, *Goodspeed*. A day later, he captured four more New England fishing schooners, burned most of them, and sent the most rickety inshore with his prisoners.[49]

Lurking square in the middle of the shipping lanes, apparently immune to the attentions of the U.S. Navy, Read was by now spreading panic along the East Coast. As Captain Maffitt had predicted, "All the merchants of New York and Boston who have . . . become princes in wealth and puffy with patriotic zeal for the subjugation of the south, will soon cry with a loud voice, peace, peace; we are becoming ruined and the country be damned."[50] Boston underwriters put a price of $10,000 on Read's head.[51] He added fake news to his asymmetric methods of warfare when he told *Goodspeed*'s crew, "Why, the other day I fired off two ships within sight of a United States gunboat, and she didn't lift a finger to stop me."[52] He intended them to spread the story when they were released, and they did, fanning the flames of panic and spurring further frantic exertions on the part of the Navy and Revenue Service.[53] Eventually over forty chase vessels were out searching for him.[54] Even the Academy's practice ships were hastily armed and sent out after him[55], and the yacht *America*, manned by mids, searched for him off the Chesapeake Capes.[56]

But Read knew his pursuers were closing in. His black-hulled clipper was becoming too well known, and he was out of ammunition for her single gun.[57] After burning two more ships and ransoming a third for a bond of $150,000, he transferred the little cannon, his crew, and his flag to a small fishing schooner, *Archer*. Then left *Tacony* burning behind them.[58]

One might think Read had raised enough hell for one cruise, but he was just getting started. Changing tactics, he sailed inshore toward Maine. On the morning of June 26, he picked up two local fishermen. Thinking their new friends "a merry crew of drunken fishermen out for a lark,"[59] they freely spilled the local news, including the fact a U.S. revenue cutter was moored in Portland harbor. Read resolved to seize it; an armed cutter would make a fine raiding cruiser.[60]

That night Savez blithely ghosted his innocent-looking ninety-ton mackerel trawler in past the guns of Fort Preble, into Portland harbor.[61] There he anchored near the Revenue Service cutter *Caleb Cushing*. After mastering it, he hoped to fire other ships in the harbor, and perhaps even burn the city.[62,63] After a prayer with his men, Read led them into two small boats, and made up on the cutter's stern. Finding most of the crew ashore on liberty, Read ironed the captain (his classmate Dudley Davenport) to the mast. Unfortunately there was no wind, the tide was against them, and it took half an hour to get the anchor up, after which the cutter became stuck in a mud bank. Finally freeing her, Read got back into the boats with his men, grabbed an oar, and began towing his latest capture out to sea.[64]

So far, so good. But the next morning the wind died again. The townspeople discovered the cutter was missing, and piled local regulars into steamers. They caught up twenty miles off Cape Elizabeth. Read prepared for battle, but discovered that though he had plenty of powder, he had no shot. (Actually there was both shot and shell aboard, but it was hidden, and Davenport refused to give up its location.)[65] Read fired the sole round shot he could discover, some iron junk, ballast stones, and (some say) a large cheese at the approaching steamers.[66] He put his prisoners in the boats, then his crew, and at last debarked himself, after setting a fuze to the magazine. Four hundred pounds of powder blew the Federal cutter to splinters, and Read was in the hands of his enemies at last.[67] "Captain Read is described by all who saw him as little more than a boy, bright-faced, alert, twenty-three years of age, rather slight, with brown moustache and whiskers and a thin, sharp face...."[68]

A "boy," perhaps, but still the enraged citizens of Portland wanted to hang the "pirate" then and there.[69] Saved by his classmates in the Union navy,[70] Read was imprisoned in Fort Preble before forwarding him to Fort

Warren, in Boston Harbor.[71] From that gray granite fortress, Read penciled a report to Secretary Mallory.[72] In the fifty-two days since leaving Florida off Brazil, he had captured or destroyed close to twenty ships, including *Cushing*, cost the Union millions of dollars, and occupied upwards of forty ships in haring after him, all without hurting a single person.

Yet even in prison, he could not rest. After less than a month, he and several other POWs discovered that they could squeeze through an eight-inch-wide musket-slit in their bathroom, if they took all their clothes off.[73] They climbed down a stone wall, and trusted themselves to floats made out of empty bottles. But hypothermia, the tide, and storm defeated them; they returned to their cells before they were missed.[74]

Read's next escape attempt was via a disused chimney. As he hid under some canvas a guard stabbed a bayonet into his leg. Read remained quiet and swam to a small sailboat, but almost bled out before he was recaptured a second time. Now he had a bad leg and had lost the sight in one eye digging out of the chimney flue. He was finally exchanged (swapped for a Union captive in Confederate custody) in October of 1864.[75]

Returned to duty, though emaciated and half blind, Read—now with a new sobriquet, "Tacony"[76]—was assigned to the defenses of Richmond on the James River. He set to work commanding two gun batteries in near-constant action near Trent's Reach.[77] He was soon bored with that, though, and applied for duty with a "torpedo boat" (submarine) being built in Charleston, SC.[78] (Perhaps *Hunley*? In which case, he was lucky to be turned down.) He next presented Secretary Mallory with a plan for an overland raid to capture a tug, rig a spar torpedo, and destroy CSS *Atlanta*, which had been captured and lay under the guns of Fortress Monroe.

On a dark night in November, 1864, Savez set out with a small party, circumnavigated the Union lines, and rowed down toward Hampton Roads. He boarded and took a sutler's schooner, then, after a short but deadly firefight, a Union tug that anchored nearby. But the rifle shots alerted other Federals. Read freed his prisoners, scuttled the tug, burned the schooner, and retreated. He tried the same type of special operation again a few weeks later. This time, he burned three more schooners and another tug, and accepted the "surrender" of some Confederate cavalry who took his blue-shirted sailors for Union troops. But once again, he couldn't break through to his intended target.[79]

Read's reconnaissances, though, had alerted him to a gap in the Federal barriers downriver. He took this intelligence to Higher, pointing out there would never be a better chance for the bottled-up fleet to strike a major blow against Grant's supply base at City Point, an action which might have broken the Union siege of Petersburg.[80] This was especially true because most of the Federal fleet was off Wilmington at the moment, reducing Fort Fisher. Only one ironclad, *Onandaga*, remained.[81]

At 6 PM on January 23, 1865, the James River Fleet began to move, with Read in charge of the command's torpedo boats. His old shipmate James Morgan encountered him that night:

"The night was very dark, and suddenly I heard a sentry challenge something in the river. I ran down to the edge of the water and arrived there just in time to see a rowboat stick her nose into the mud at my very feet, and was much surprised to see my old shipmate, "Savez" Read, step ashore. He was in a jolly mood, as he told me that our ironclads would follow him in a couple of hours, and that he was going ahead to cut the boom so that they could pass on and destroy City Point. "And now, youngster," he said, "you fellows make those guns of yours hum when the 'Yanks' open, and mind that you don't shoot too low, for I will be down there in the middle of the river." And then he put his hand affectionately on my shoulder and added:

"Jimmie, it's going to be a great night; I only wish you could go with me: a sailor has no business on shore, anyway." And laughing he stepped back into his boat and shoved out into the stream."[82]

The fleet took heavy fire in Trent's Reach and several ships ran aground. Read went forward in *Scorpion* to take soundings and inspect the Federal obstructions, under heavy fire.[83] He reported back they could be cleared, and returned to do so. Shortly thereafter *Fredericksburg* bulled through. But the strongest ironclad, and the flagship, *Virginia II*, had gone solidly aground, along with others. Read returned to help, but promptly went aground himself in the shallow river at ebbing tide. Exposed at dawn to Federal batteries on the heights, the fleet took heavy damage before another tide allowed them to retreat upriver.

Though they tried again the next night, they were no more successful, and abandoned the effort. The final chance to break Grant's investment of Richmond had failed.[84] On February 23, Read led one final whaleboat-and-torpedo expedition to "Blow up the Federal iron-clads, clear a passage for our fleet and force the abandonment of City Point, or compel Grant to fall back or bring his supplies from Norfolk."[85] But he was betrayed by deserters, and escaped capture only by wading across the frozen Appomattox River.[86]

By March, 1865, Richmond's end was in sight. But Read was back on Stephen Mallory's desk, pushing a scheme to fit out a new raider in England. Mallory switched his firebrand lieutenant to an existing ship. He gave Read sealed orders to report to Shreveport, Louisiana, to take over *William H. Webb*, a fast, powerful 655-ton sidewheeler "cottonclad" ram[87] that had seen action near Vicksburg. Read crossed nearly two thousand miles of mostly federal-occupied territory, evading patrols, to report to General Kirby Smith.

Webb had to be manned, gunned, fitted out, and extracted through three hundred miles of Federal-occupied river, but by this time Read knew how. The Cause was in its last hours, but he drove ahead, intending to make Cuba his home base and recommence the guerre de course he'd been so successful at before.

Getting underway on April 16, 1865, he fought and bluffed his way down the Red River, then the Mississippi, running a gauntlet of enemy fleets and shore batteries. Lee had surrendered. Lincoln was dead. But Read was charging south with a spar torpedo hanging off his bow, dodging between monitors, cutting telegraph wires as he went. His crew lounged on deck in captured Yankee uniforms, imitating a Federal supply ship. Read ran through New Orleans so fast and unexpectedly that "It was two hours and a half before Admiral Thatcher's squadron off that city recovered sufficiently to start pursuit."[88] But one ship, *Lackawanna*, opened fire, alerted by a sailor who recognized *Webb*. The shot cut loose Read's spar torpedo, which now dangled over the bow. With a live detonator, it could neither be pulled back aboard nor safely cut loose. Read stopped the ship and "sprang forward and cut the explosive. 'I got it! Full speed ahead, Charles called back (to his pilot)."[89]

U.S. sloop-of-war Richmond, *of Farragut's fleet*

Burning of CSS Webb *below New Orleans, April 24, 1865*

Below the city though, Read encountered an obstacle even he couldn't circumvent, or trick his way through. USS *Richmond*, coming from sea, ". . . took such position in the channel that the *Webb* would have had to pass close under the powerful broadsides of the man-of-war. Read saw that he must lose his ship; and, preferring to save his and his men's lives, he took the bank, setting fire to his grounded vessel and fleeing into the swamp."[90] Read surrendered his sword to his classmate Winfield Schley, now captain of a Union warship.[91]

After another spell in Fort Warren, Read was amnestied. Though he was one of the most famous naval officers of the age, his valor had been employed in a losing cause. His military career was over forever. And he was only twenty-five.

After his release, Read headed for Liverpool to look for a job as a merchant captain. But his pardon came through, and he returned to New Orleans. Borrowing some money from James Morris Morgan, with whom he had served aboard *McRae*,[92] he tried captaining a fruit schooner, but this venture failed.[93] Shortly thereafter, Read's 1860 classmate E. G. Read (no relation) searched him out with a shady scheme. They would buy a surplus gunboat in New York, repair it, vanish out to sea, and sell it to the president of Colombia, which was under an arms embargo just then.

Read delivered the ship and collected his pay. Then he boarded it by night, much as he had *Caleb Cushing*, stole it back, and sold it a second time, this time to the rebel faction.[94] Compounding his chicanery, he then offered to steal it yet again, returning it to the original purchaser for another payment. He was informed that if he was caught in Colombia again, he would be hanged.[95]

Returning to the U.S. once more, Read married Rozinda at last in 1867. He settled in as a loyal citizen, father, and river pilot, and served as harbormaster and then president of the harbormaster's board for New Orleans.[96] He died in Meridian, of Bright's disease and pneumonia in 1890, only forty-nine.[97]

Though the Confederacy authorized Medals of Honor, none was issued during the war. Read was, however, the subject of a joint resolution of the Confederate Congress thanking him for his part in the Battle of New Orleans.[98] The Sons of Confederate Veterans awarded him the Medal of Honor in 1979 with the following citation:

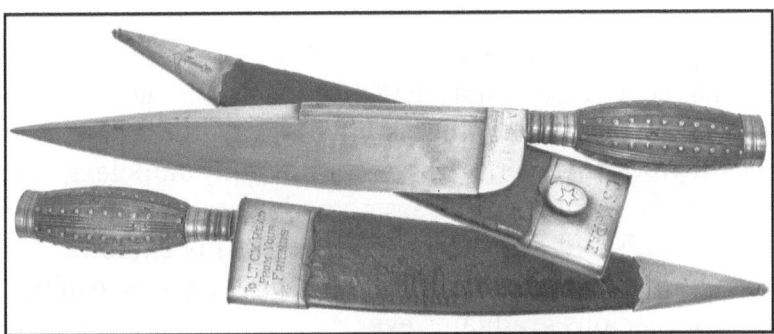

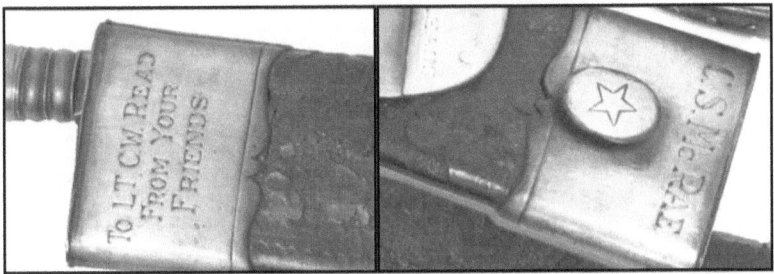

Photo credit: Janet and Bedford Hayes Collection

"Armed with but a single howitzer... Lt. Read embarked on a mission to wreak havoc on the commercial shipping interests of the enemy. In a series of bold attacks marked by audacity and novel self-reliance, Lt. Read successfully captured 21 vessels bound to or from enemy ports without the loss of a man. Twice he moved his command to capture vessels in an effort to confuse the enemy search force of more than 40 ships. Learning that the revenue cutter Caleb Cushing was anchored at Portland, Maine, Lt. Read entered the harbor and under the cover of night, quietly seized the cutter and cleared the harbor before dawn dispite the disadvantage of tide and light wind. Pursued by an overwhelming enemy force, Lt. Read kept his guns firing until his ammunition was exhausted. Realizing that further resistance was futile, Lt. Read ordered his men into lifeboats and then fired the Caleb Cushing before surrendering to the enemy, having spectacularly disrupted commercial shipping along the Atlantic coast of the United States."

Daring, resourceful, aggressive, yet mindful of the human cost of war, Charles W. Read would have made his mark in whatever conflict came his way. His tragic flaw was loyalty to a cause that was not only hopeless, but ultimately ignoble. His classmates became admirals; he finished his life as a harbormaster. Yet his exploits still resound, and his memory still merits honor, as one of our most audacious graduates ever.

Savez Read's grave marker

4
THE MID WHO MOVED THE WORLD: FRANK JULIAN SPRAGUE, CLASS OF 1878

When you push the button on an elevator, take a subway, electric train, or light rail; drive a Prius or a Tesla, step onto an escalator, serve in a ship with electric drive . . . you can thank a brilliant but unjustly forgotten engineer.

It's a strange oversight, that no Academy engineering building, program, lecture series, or award is named after Frank Julian Sprague.[1] That no plebe was ever required to spout a memorized encomium to him. Indeed, most readers are probably asking themselves: Who the hell is Frank Sprague? And what's he got to do with my life?

The answer may surprise you.

Sprague's name belongs among those of the greatest inventors of the nineteenth and early twentieth centuries, the heroic age of electrical engineering that included Oersted, Henry, Faraday, Morse, Tesla, Sperry, Maxim, Westinghouse, and Marconi. In his time he was ranked with Thomas Edison and Alexander Graham Bell.[2] He worked with some of them—and one in particular appropriated, rebranded, and buried much of his life's work.

Which is why today Sprague is hardly remembered.

Let's see if we can administer a small measure of historical justice in these pages.

Sprague was born in Milford, Connecticut, in 1857,[3] descended from a Devonshireman who left England in 1628. His father David was a plant superintendent in a hat factory in Milford, which was then a farming and shipbuilding community just turning to manufacturing.[4]

The boy was uprooted when his mother died in 1866. His father left town for opportunities farther west, parking Frank and younger brother Charles with their maiden aunt Elvira Ann ("Aunt Ann", an intelligent and respected career schoolteacher) in North Adams, Massachusetts.[5] North Adams was also rapidly industrializing, with textile manufacturing, cotton mills, canals, tunnels, and railroads. Frank grew up surrounded by rapid change, hard work, and American progress. Old newspapers refer to his boyhood mischievousness and tendency to play pranks.[6] He attended Drury High School and excelled in mathematics and science.[7]

Few young men and even fewer women went to college then, and Sprague had no funds to do so. However, encouraged by Drury's principal, he decided to try for West Point—then the leading civil engineering school in the country—where he could receive an excellent education for free.[8] But when he showed up for his four-day entrance exam, he found it was for the Naval Academy instead. Displaying excellent judgment, Sprague took it anyway and placed first of thirteen candidates.[9] He left for Annapolis with a borrowed $400 in his pocket.[10]

His entering class numbered 104 that summer of 1874.[11] Although the Navy as a whole was in a postwar ebb, that same year saw the arrival of Rear Admiral C.R.P. Rodgers as Superintendent. Rodgers "oversaw an overhaul of the curriculum, including adding upper-level electives in mathematics, mechanics, physics, and chemistry."[12] In fact, in 1878 the Paris Universal Exposition awarded

USNA the *Diplôme de Medaille D'Or* for its excellence in engineering education.[13] The skinny, sandy-haired, high-voiced seventeen-year-old had the great good fortune to arrive at exactly the right place at exactly the right time.[14]

Sprague was later remembered for a scrap at the Academy, which apparently was more indulgent toward interpersonal violence than it is today. The New Englander stood out by sticking up for a black midshipman, was sent to Coventry (shunned) for doing so, and got in a fight over it.[15] A classmate later wrote to him, "You have not the pug nose of a fistic champion, were a sorry sight after the battle, but you licked your man."[16] However, his ostracism blew over, and Sprague distinguished himself academically. He became especially interested in electricity, and "was encouraged in this by a great teacher, Lieutenant Commander William T. Sampson, then head of the Department of Physics."[17] As a first classman he wrote to Thomas Edison, asking for the loan of some telephone apparatus, the first hint of a relationship that would have far-reaching effects.[18] He also observed, or at least caught glimpses of, the speed-of-light experiments of Albert A. Michelson, then an instructor in physics and chemistry.[19] [20]

He left USNA as a Passed Midshipman in June, 1878, seventh in a graduating class of fifty, with honors in math, chemistry, and physics.[21] "Sprague came out of Annapolis equipped with both a fundamental grasp of scientific electrical theory (circa 1878) and a resourceful capability for 'craft knowledge' (in the sense of hands-on trial and error) as a means of working toward technical solutions."[22]

On his post-graduation leave, Sprague stopped at Menlo Park for an encounter that would change his life, and not always for the better. Thomas Alva Edison was already world famous for the quadruplex telegraph, carbon telephone, incandescent light, phonograph, and other inventions, and was starting work on electrical distribution systems. The "Wizard of Menlo Park" welcomed the young officer and took him on a personal tour of the lab.[23] Sprague must have thought he'd been taken up to heaven.

The Navy, though, had other ideas. He was ordered to USS *Richmond*, then deploying to the Asiatic station. Launched in 1860, his first ship was a relic of the age of sail, a wooden steam sloop that had fought at New Orleans, Vicksburg, Port Hudson, and Mobile Bay. En route to the Pacific, *Richmond* called at Gibraltar, Naples, and Singapore. As the flagship, Richmond visited Japan, China, and the Philippines, at least giving her junior officers a look at the world, if not challenging them technologically.[24] Sprague used the time for thought. He filled notebooks with dozens of ideas for inventions, primarily electrical ones, such as self-regulating electric lights, ice machines, make-and-break armatures, turbine governors, "vibratory telephonic octoplexes," printing telegraphs, and most notably *constant force motors*.[25] But he must have felt increasingly frustrated at being so far from the exploding technological scene on the other side of the world, in Europe and the United States.

His next duties brought him closer to the action. After failing to pass his two-year exam back at Annapolis (he flunked Seamanship), but passing on his second try[26] and being promoted to ensign, Sprague worked at the Brooklyn Navy yard on arc lighting and dynamos before being assigned as gunnery officer on the training ship *Minnesota*. He tried to wire *Minnesota* for electric lighting, but couldn't get satisfactory equipment. When the frigate was ordered to Newport, though, Sprague linked up with Professor Moses Farmer, then in charge of development at the Torpedo Station, a center for naval electrical development.[27] Farmer had developed arc lamps and dynamos, and had even built a little two-man electric car years before, supplying current through the rails for the first time instead of by an onboard battery.[28]

There, Sprague began work on an idea he'd been considering for a long time.[29]

It was an electric motor, but of a new type.

The constant-speed Sprague motor, or "Inverted Type Dynamo-Electric Machine" was the result of these early experiments. It made two radical departures from the then-classical motor/dynamo design as developed by Faraday, Henry, Pixii, Gramme, Davenport, and others.[30] [31] In those designs, familiar today in toy electric motor kits, the armature rotated between fixed coils that generated a magnetic field. Sprague moved the field magnet inside the rotating armature and enclosed it with iron wire and projecting iron ribs. He then outlined a way to shunt or switch between different combinations of field and armature components.[32] Together,

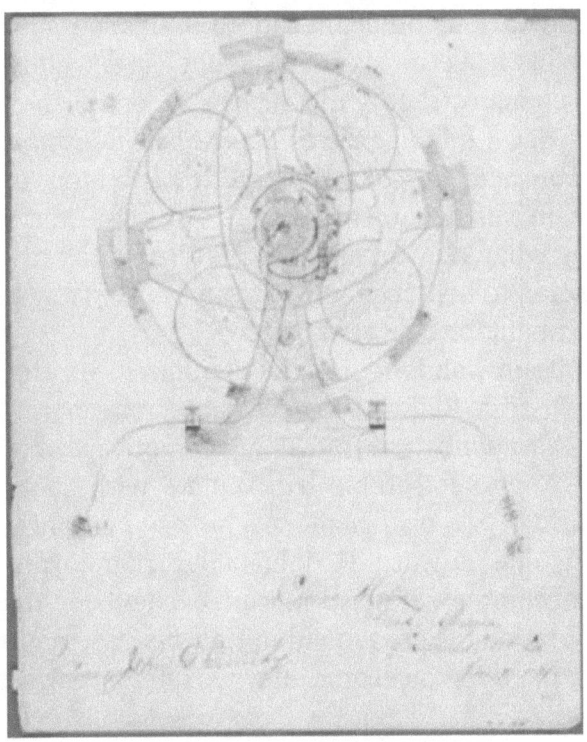

Design for the constant-speed Sprague motor

the innovations promised increased simplicity, efficiency, and flexibility both for dynamos (DC generators) and motors.

Sprague wanted to go to a planned electrical exhibition in France, and Farmer interceded with the SecNav to get him a Med-bound ship. He was assigned to USS *Lancaster*, heading east for the Mediterranean Squadron.[33]

Alas, *Lancaster* got to Europe too late for the French exhibition. But Sprague wangled three month's leave to attend another major exhibition at the Crystal Palace in London in 1882. He arrived in England with just twenty dollars in his pocket, but quickly found employment.[34] Young as he was, he introduced himself, made a good impression,[35] and was named secretary of the jury of awards under Lord Kelvin, involved in testing and ranking generators, lights, and various other inventions.[36] He also managed to overstay his leave by six months, but "saved himself with a comprehensive report,"[37] which was republished by the Office of Naval Intelligence as *Report on the Exhibits at the Crystal Palace Electrical Exhibition, 1882*.[38] (More or less as an aside, in this document Sprague also came very close to inventing what later became the Diesel engine.) It won him a promotion to lieutenant and quietly vacated any prospect of punishment for being AWOL.[39]

While in London, Sprague commuted on the Metropolitan District Railway, a new subterranean railway, or "subway," drawn by coal-burning locomotives, which made for a dirty and miserable trip replete with soot, smoke, and sparks.[40] He also made useful acquaintances there, including one Edward H. Johnson, a flamboyant young Edison promoter who also recruited talent for the Menlo Park operation.[41] Sprague submitted his resignation from the Navy in 1883, to become effective after a year's leave.[42]

Sprague had already done some research on lighting circuits before leaving England. Neither he nor Edison had invented them, but they became the basis for the 220/110 volt three-wire DC system Edison was preparing. On arriving in Menlo Park, the 25-year-old Sprague was promptly assigned to install the first overhead three-wire electrical system in the United States, at Sunbury, Pennsylvania.[43] Edison had designed the laydown and gauge of its wiring by means of a scale model, with simulacra of houses and streets. Sprague quickly saw that he could do this faster and more accurately with mathematics, saving time and expense over Edison's cumbersome experimental approach.[44] (Nikola Tesla, who was also working for 'The Chief' at the same time, also noted Edison's weakness in theory and sometimes bullheaded dependence on empirical methods.[45]) Sprague patented this algorithm in 1885, but assigned the patent to his employer.[46]

Sprague only worked directly for Edison for about a year. He wanted to develop power transmission and electric motors, not residential lighting systems, and had in fact been working on motor designs at night and on weekends.[47] But rather than reassign him, Edison fired him. *Scientific American* wrote in 1911, "Here was the parting of the ways. Edison has always wanted associates who were all his own, on whom his dominating personality could be impressed, who would help carry out his prolific ideas and not incidental inventions of theirs."[48]

The now self-employed Sprague resumed the work he'd begun at the Torpedo Station. Up until then, electric motors had been expensive, undependable, and weak. They slowed down or stalled when loads were imposed, and each one required an attending electrician to adjust the brushes during operation to prevent sparking and fires.[49] Sprague's would be an enormous step forward. Working at incredible speed, he had operating models of his improved motor designed, built, and ready for the International Electrical Exhibition that opened in Philadelphia in September, 1883. (One of these prototypes is on display, still running, at the Shore Line Trolley Museum in East Haven, Connecticut.) "Sprague's motors displayed at the exhibition were well designed, incorporating several new features, and their merits were quickly recognized. The motor was virtually non-sparking, and operated at a constant speed, regardless of load." [50]

This last characteristic may have been the most notable feature of his invention. One reference defines what are now called "compound-wound" motors this way: "The DC compound motor is a combination of the series motor and the shunt motor. It has a series field winding that is connected in series with the armature and a shunt field that is in parallel with the armature. The combination of series and shunt winding allows the motor to have the torque characteristics of the series motor and the regulated speed characteristics of the shunt motor."[51] As Sprague also saw around this time, it could also be used to *regenerate*— to turn excess mechanical energy back into electricity—and could thus be useful for braking.

His essential theoretical insight was, as he put it, ". . . to maintain the speed of a constant potential motor, constant under varying loads, when the load increases, the field should be weakened; and when the load is decreased, the field should be strengthened." Recast in the form of equations from which nearly any type of motor or dynamo could be designed, these became known as the Sprague Laws.[52]

Even Edison complimented him, calling his invention "the only true motor."[53] Sprague incorporated the Sprague Electric Railway and Motor Company (SERM) in 1884, but it existed largely on paper.[54] Manufacture of his motors was subcontracted to the Edison Machine Works and sold via Edison sales channels. The new motors sold briskly, powering pumps, blowers, lathes, looms, and hoists.[55] Sprague leased his own factory eventually, but he was already moving on to a related question he'd been cogitating on since London.

How could electricity be used to more safely and efficiently move human beings along rails?

By the end of the 1870s horses, including omnibus horsecars, were a major problem in cities. "At a rate of 22 pounds per horse per day, equine manure added up to millions of pounds each day and over a 100,000 tons per year (not to mention around 10 million gallons of urine). Per one observer at the time, the streets were "literally carpeted with a warm, brown matting . . . smelling to heaven."[56] Elevated steam railways, the "El," added smoke, cinders, ash, and steam to the redolence. Sprague was determined to deprive billions of flies of their livelihood.

Around this time, he also deprived New Orleans of one of its most beautiful belles by marrying Mary Keatinge, whom he'd met on a short vacation trip to that city, in 1885.[57]

Sprague devoted 1885-86 to developing electric propulsion for use in New York City elevated trains. His approach was radical. He proposed to put motors beneath *each car,* instead of having a long train towed by a dedicated locomotive. Their motors would turn into generators when the car had to brake or go downhill, returning power to the system. He also developed a brilliantly innovative three-point "wheelbarrow" mounting that flexibly coupled the motor to the drive wheels, an advance that would become universally adopted for electric railways.[58]

Sprague built a test car with the new motor and mounting, running it on a short track fitted with a third rail at a sugar refinery at East 24th Street. He had railroad tycoon and Manhattan Elevated owner Jay Gould along for a ride when "the fuse blew with a startling flash, followed by Mr. Gould's attempt to jump off the car despite the remonstrance of all."[59] But in the end the El owners decided converting to electricity would entail too much recapitalization,[60] and probably Gould's public humiliation didn't help either.[61]

Sprague had to cut his losses. But as usual, he had another idea in mind.

In 1887, he began work on a "street railway"—what we would today call a light rail system—for the Richmond, Virginia Union Passenger Railway. [62] At the time, there were nine streetcar installations in Europe and ten in the US,[63] but their arrangements varied widely and all were unsatisfactory for various reasons.[64] Richmond mayor John S. Wise wanted an improved street rail system to tie outlying Reform Republican wards into the city. A New York investor had helped secure financing for the project, and he recommended Sprague for the job.

But it would be difficult. Not only would Sprague have to design and build all new motors and cars, he had to build an electrical plant to power them. He had to supply high voltages to a moving vehicle without electrocuting pedestrians. Even worse, Richmond's hills were very steep; the town was known as a "horse-killer." [65] The long climb at Franklin Street reached a grade of up to 1 foot in 10, a daunting prospect for electrical traction. But the line had to be up and running in *ninety days*. The contract called for Sprague to front all expenses, to be repaid only after 60 days of satisfactory operation.[66] [67] "He was uneasily aware that he had undertaken to build more electric cars than there then were in the entire world, and that his, unlike the others, would really have to work."[68] And just as he was ready to start, he caught typhoid and was down and out for nine weeks.[69]

"Failure in Richmond meant blasted hopes and financial ruin," he later wrote.[70] And when he finally got there, after recovering, he found the newly laid tracks were totally inadequate. But he'd assembled a can-do team, made up largely of West Pointers and Naval Academy grads. These included Oscar Crosby, USMA 1882, and S. Dana Greene, USNA 1883 (son of the executive officer of USS *Monitor*, Samuel Greene USNA 1859).[71] They got an experimental car with two 7.5 horsepower motors up Franklin Hill in November, 1887, though it overheated, derailed, and had to be pulled back to its shed by mules. Further expensive and time-consuming development followed, of the motors, of the gearing, of grounding, of the commutators, insulation, track, and of Sprague's novel overhead wire arrangement. (The 'trolley' is actually the spring-loaded, wheeled device that assures continuous contact with the powering wire, invented during this time frame by one of Sprague's employees. It became the name for the "trolley car" by extension.) The phone company and gas company sued them. Lightning burned out the lines, forcing Sprague to invent the lightning arrestor.[72] They didn't meet the ninety-day window and had to accept a reduced payment. Sprague had to take out personal loans, putting all his capital earned from his motor business at risk.[73] He wrote, "I am completely overwhelmed with work, so much so that I hardly know if I stand on my head or my heels."[74] In the midst of this, the first troubles with his marriage to Mary were beginning to surface.[75]

But by January they had nine cars operating. More teething problems bedeviled the engineers, but by May 4 they had thirty cars up and something resembling a citywide system going. On May 15 Sprague was informed he'd met the contract specs and would be paid. But by then he'd lost over $75,000 on the contract.[76]

However, his loss proved a gain. Richmond was the first commercially successful large-scale trolley line in the world, and SERM powered ahead in the rapidly expanding electric traction market. "Within two years of its opening, 110 electric railroads using Sprague's system were built or under contract, including systems in Italy and Germany."[77] Boston developer Henry M. Whitney, owner of the biggest transit system in the world, the West End Street Railway, had been searching for an answer to that city's incredible congestion (as well as a way to increase the value of his suburban landholdings.) Whitney came to Richmond to see for himself. When Sprague lined up 22 cars at the bottom of Church Hill, and sent them all up at once, Whitney was convinced.[78] He ordered the world's second electric rail system, and "... others followed around the world, with 20,000 miles of streetcar tracks laid in the United States alone by 1905. His design served as the basis for a variety of systems later built by competitors such as Thomas-Houston, General Electric and Westinghouse."[79] Sprague and his team had perfected efficient, fast, clean, cheap, and dependable transportation for millions, not just removing the gritty, disease-promoting muck of manure and cinders from the streets, but permitting cities to sprawl to a degree never before possible.

Yet even in triumph, Sprague was being stalked by a predator. Despite its technological lead, his company was undercapitalized and overextended. As sales flooded in, Sprague's ownership was diluted by stock purchases by Edison General Electric, which wanted in on the traction market. Trolleys quickly transformed the face of America, accelerating the growth of cities, making possible "ribbon communities" along lines, and offering new work opportunities to young people and women.[80] But by 1890, SERM had been absorbed by EGE, and the inventor was relegated to being a consultant, isolated within the company he'd created.[81] Not only was he out, from then forward his name was removed from the motors and other equipment he had designed.[82][83][84] His inventions bore the name "Edison" now, and Sprague wrote a bitter letter of resignation from the company he'd founded.[85]

But once more, he reinvented himself. Free again, he briefly investigated applying electricity to naval gun pointing and training, but eventually turned to locomotives, subways, and elevators. Working with Dr. Louis Duncan (USNA 1880, PhD. Johns Hopkins 1887), they developed and built the heaviest, most powerful electric locomotive yet seen for Henry Villard of the Northern Pacific.[86] The sixty-ton behemoth worked perfectly, but the road went bankrupt and it was never put into service. And in New York, the spark-shy Jay Gould was still stonewalling electrification of the El.

So Sprague turned to elevators.

In the late 1800s, steel-framed construction was making it possible to extend buildings upward. In 1885 the ten-story-high, metal-framed Home Insurance Company Building went up in Chicago. Elevators were usually driven by steam, with power transmitted either by rope or hydraulically-driven pistons. But as story piled on story, pistons became impractical.

The first lumbering electric elevators, hoisted by gears on toothed racks, were built by Siemens in Germany. Sprague began with a sheaved cable-hoisted design outlined by a young engineer named Charles Pratt, and added the motors and control mechanisms he'd developed for his trolley ventures. Giving Pratt appropriate credit, Sprague partnered with him and his old friend Edward Johnson. By Fall of 1891 their first installation was in operation, in the Grand Hotel at 31st and Broadway.[87]

This too required tinkering and reworking, as did the intrepid inventors' corporate structure. In 1892 the three pooled patents and holdings to form the Sprague Electric Elevator Company.

In what seems like karmic payback, that same year a J.P. Morgan-engineered coup took Edison's name off what now became just plain General Electric.[88]

Once again, though, Sprague and his partners faced tough competition. Not only that, the Panic of 1893 struck terror into investors and builders alike. Once again, Sprague had to commit his personal fortune in order to move forward. This time it was to provide six direct-operated electric elevators with multiplying sheaves, electric braking, and distant pilot control for the fourteen-story Postal Telegraph Building at Murray and Broadway. (Around this time, he also invented the push-button control, first installed in an elevator in Massachusetts.[89] [90]) Sprague not only guaranteed that they worked, but promised that if they didn't he would install hydraulic elevators at his own expense.[91] As he had in Richmond, he was staking everything on his team, an untried technology, and his own innovation.

And it nearly killed him. The contacts on a controller welded shut when he and most of his crew were in the elevator testing it *before any of the safeties had been installed.* It shot toward the overhead sheaves at top speed, and only an alert employee in the basement saved them by slamming the master switch to off.[92]

"Sprague never had to make good his guarantee. He ran four of his elevators at 325 feet per minute with 2500-pound loads and the other two at 400 feet per minute with 1800-pound loads . . . Sprague's company developed floor control, automatic elevators, acceleration control of car safeties and a number of freight elevators . . . He connected individual control circuits to a master control switch to operate all the elevators at the same time. Sprague's idea worked. However, the cars' operations were hardly synchronous, and he continued to ponder the problem."[93]

Meanwhile he continued to tinker with the Grand Hotel elevator, and one night wondered what would happen if he started all six elevators at once. After some thought, he linked together the pilot motors, which drove the contacts that started the larger hoisting motors. When he closed the master switch, all six elevators rose in perfect synchronization. Recalling this experiment several years later, he ". . . realized he had not only discovered a means to more rapidly, precisely and flexibly controlling a car, but also a means of controlling cars in groups."[94]

First two Multiple Unit Cars, 1897

Once again, Sprague tasted success. He provided six hundred elevators for the tallest buildings in New York,[95] built in a 30,000-square-foot, 850-employee factory in Watsessing, New York, where he also produced armored cable and the first automatic watertight doors.[96] Then, with the technology solved, he sold his elevator operation and patents to Otis.[97] Since that problem was solved, the inventor felt he could step away. And turn his hand to the next invention: one he felt might be the most impressive and profitable of all.

Despite his involvement in trolleys, electric locomotives, and then elevators, Sprague had never stopped pushing for subway/light rail/elevated electrification. By 1896, his midnight experiments with elevators had led him to what he called the Multiple Unit Control System.

Instead of all cars being pulled by a (prohibitively heavy) locomotive, each was motorized, like a trolley car. But unlike a trolley, the lead ("pilot") unit in a train could command them all, starting, stopping, and varying speed simultaneously by means of relays. Any unit could be a pilot car, could go in either direction, and cars could be added or subtracted to a train at will—significantly increasing a line's flexibility to adapt to rush-hour demands.[98] The distributed load was easier on elevated structures, and having many powered wheels instead of a few gave better traction and faster acceleration than locomotives could ever achieve.

Sprague's first installation of this "M. U." system was for the elevated railroads in Chicago. Once again the contract was intimidating. In only two and a half months, he had to turn a paper concept into operating machinery, and guarantee performance with a $100,000 personal bond. If all 120 cars could not perform perfectly he would not be paid at all.[99] Not only that, just then he had major ongoing elevator installations in London, and was on crutches, having broken both legs in a fall from a scaffold in the Waldorf-Astoria.[100] [101]

Long hours of work ensued, in the midst of which Sprague's marriage to Mary ended in a quiet divorce. His team put their first cars on the Loop in late 1897 with the West Side Railroad and completed the South Side contract in summer of 1898, though at a loss of some $51,000.[102] Sprague's control system worked so perfectly his ten-year-old son Desmond drove a train in demonstrations, and two cars could be operated together even without the usual coupling links and pins.[103] [104] Even more appealing for the transit owners, the South Side reported that their profits tripled after sidelining steam locomotives.[105] An installation in Brooklyn came next, and also proved a success (after overcoming burning fuses and overheating motors).[106] The MU system quickly became industry standard, not only in the US but overseas, furnished by Sprague-Thomson-Houston in the UK and the Societe Francaise Sprague in France and Germany.

The Sprague shops in Bloomfield, NJ

Unfortunately, Sprague was also locking horns with the "Electric Trust"—Westinghouse and GE, the latter deploying his earlier Edison-assigned motor patents against him while they reverse-engineered a rival control system. A burglar stole the MU plans from his office; he suspected an agent for GE.[107] Sprague was trying to put together a third titan, a vertically-integrated industrial platform that could take on the two bigger players, but that effort cratered when his stockholders buckled and GE bought him out yet again.

Once more he was merely a consultant, but this time, at least, he ended up with the equivalent of fifty million dollars in today's currency and continuing royalties on his M. U. system. As well as an agreement that this time, his name would remain on the "Sprague General Electric Multiple-Unit Control."[108]

Sprague had written to Theodore Roosevelt, then Assistant SecNav, to volunteer for sea service in the Spanish-American War. But shortly afterward he was nearly blinded in a factory accident, and the war ended before he could make the physical.[109] In 1899 he met and married Harriet Chapman Jones, of Connecticut, and over the years they had three children. The family divided their time between a townhouse on the West Side and a palatial home in Sharon, Connecticut. Harriet was literary, a rare-book collector and Walt Whitman expert. She introduced her husband to writers and dramatists, including William Dean Howells, Booth Tarkington, Emil Ludwig, Albert Bigelow Paine, and the father of her friend Clara Clemens, Mark Twain.[110]

After a horrific accident in 1902, when an engineer missed a smoke-obscured signal and two trains collided in the Park Avenue tunnel—killing fifteen—the New York legislature ordered the NY Central to electrify. That order coincided with the line's burning need to improve and update the circa-1870, and increasingly crowded and inadequate, Grand Central Depot and its approaches,[111] Sprague served with Louis Duncan on the Commission for Terminal Electrification from 1896 to 1900,[112] and the specifications were largely those he recommended. Providing disinterested technical advice, he staunchly defended the best solutions against political and special interests. In 1908 Sprague and William Wilgus, the civil engineer in charge of the reconstruction, patented a protected third-rail system. It is still in place on the Metro-North in the Bronx and part of the Hudson Valley,[113] and was widely adopted elsewhere as well. Today's Grand Central Terminal stands as a monument to their work.

With the onset of World War One in Europe, Edison suggested that America mobilize its engineering and scientific talent to meet the impending challenge.[114] SecNav Josephus Daniels organized the Naval Consulting Board in 1915, with Edison as overall chair. After a month's cruise with the Atlantic Fleet, Sprague turned in a scathing report on the Fleet's balance in terms of ship types and readiness for major war.[115] He headed up committees for shipbuilding and electricity, and served throughout the war on committees for submarines, explosives, ordnance, and "special problems." He devoted particular attention to armor-piercing shells. With his and Mary's son Desmond, he designed and personally tested what we today know as the delayed-action armor-penetrating fuze, along with pressure-sensitive firing mechanisms for depth charges.[116] [117] One of the Board's most important accomplishments was the founding of the Naval Research Laboratory.[118]

Following the war, Sprague garnered so many industry awards and recognitions the list would be too long to include here. Still, he carried on inventing. Concerned with the growing number of railway accidents, he established the Sprague Safety Control and Signal Corporation to develop a robust, durable, and fail-safe system of sensors and magnetic control points. Though carefully human factor engineered not to interfere with normal operation, the Automatic Train Control system would automatically stop a train if danger existed, independently of the engineer.[119] [120]

Sprague in 1921

Though tested by the New York Central, it wasn't adopted until the 1920's, and even then not everywhere. "Finally, nearly 80 years later, after another deadly collision, in 2008 President George W. Bush signed a railroad safety act requiring an updated version of Sprague's ATC on intercity commuter and Class I railroads by 2015. How many lives were lost unnecessarily during those 80 years is unknown."[121]

At age seventy, he patented a system for running multiple elevators in the same shaft. He licensed this design to Westinghouse. In the 1930's he also developed aircraft autopilots and a forerunner of what we would now call a digital sign.

Frank Sprague died in 1934 of pneumonia, angry on his death bed that doctors would not let his lab assistants in to review some circuit modifications he'd just thought of.[122]

He was buried in Arlington National Cemetery. In 1947 Harriet published *Frank J. Sprague and the Edison Myth*, which debunked the then-ubiquitous fable that Edison had invented practically everything electrical, including innovations that the historical record clearly showed were not his own.[123] She lived on until 1969.[124]

Following in his father's footsteps, their oldest son, Bob, graduated from USNA in 1919 with his war-accelerated class of 1921. He earned a BS from the Naval Postgraduate School and an MS from MIT. In 1926, while working in the Charlestown Naval Shipyard on the propulsion equipment for USS *Lexington*, Bob founded the Sprague Electric Company,[125] manufacturing variable condensers and fixed capacitors, important components for the burgeoning electronics industry.[126] Located in a sprawling complex in North Adams, where Frank had grown up in his aunt's home so many years before, "Sprague physicists, chemists, electrical engineers, and skilled technicians were called upon by the U.S. government during World War II to design and manufacture crucial components of some of its most advanced high-tech weapons systems, including the atomic bomb."[127] The Black Beauty and Orange Drop, Sprague Electric capacitor designs, are still being sold. But the plant in North Adams closed circa 1985 as the tsunami of Asian components capsized the domestic industry. Its buildings now house the Massachusetts Museum of Contemporary Art.

Frank Julian Sprague wasn't a hero in the *military* sense. But he led in a related field, one in which Naval Academy graduates excelled and pioneered throughout the late nineteenth century and on into the twentieth.

Sprague's work shaped cities around the world, both horizontally and vertically. His work on feedback and self-regulating systems, starting with the constant-speed motor and continuing with the Multiple Unit system, automatic train control, and other inventions, looked forward to self-regulating devices, automation, and the autonomous systems emerging today.

Sprague was the very model of the heroic inventor, who conceives new visions and perseveres in the face of failure after failure. He was the classic entrepreneur, who shoves all his winnings back on the table, again and again, to further disruptive new technologies against daunting odds.

Who the hell was Frank Julian Sprague?

Now you know.

5

PHILO N. MCGIFFIN, CLASS OF 1882: AMERICAN MANDARIN

We are what we pretend to be; so we must be careful what we pretend to be. —Kurt Vonnegut, *Mother Night*

Early in the morning of February 11, 1897, a 37-year-old man lay in pain in a Manhattan hospital, holding the revolver he'd smuggled into his private room.[1] He'd already penned a lighthearted note:

> ... with apologies to Miss Phelps for the row—
> it is the way that all guns have. Au revoir.

In those last moments, Philo Norton McGiffin must have wondered how history would record his life.

Every midshipman knows Philo's name. He's the Rebel, the Prankster, the mythical Trickster. But what do we really "know"? That he was a mid—fired the cannons—got a lot of demerits—graduated, but wasn't commissioned—went to China—was wounded in battle—shot himself in a hospital.

But he's no myth. Yellowed clippings testify to his service, suffering, and death. Nimitz Library holds his conduct records. The souvenirs he bought, his Academy uniforms, the flag he fought beneath, even the bloody trousers he wore at the Yalu can still be seen. Beyond the recycled stories, past legend and exaggeration, we can still discover the real Philo Norton McGiffin.

Philo was born in the farming community of Washington, Pennsylvania, on Dec. 13, 1860, in the back room of the county jail[2]—his father, Norton, was the sheriff. The McGiffins were proudly descended from the warlike MacGregors and MacAlpines,[3] and family lore recorded pirates and soldiers of fortune among their ancestors.[4] Philo's great-grandfather Nathaniel emigrated from Scotland, wintered at Valley Forge and fought at Brandywine and Trenton[5] before being granted land in southwestern Pennsylvania. Washington County had an insurrectionary past too—the Whiskey Rebellion had started there in 1794, and took 13,000 troops to put down.[6]

The Mexican War was unpopular in the North, but Norton was quick to volunteer. He went as a private

Sheriff's House in Washington PA, where Philo was born

in Company K of the First United States Artillery. The regiment saw fierce fighting, including the landing at Vera Cruz, the siege of Puebla, and the assault on Chapultepec.[7] After the war the county awarded Norton a set of pistols to honor his gallantry,[8] and elected him sheriff.[9]

Philo was born four months before Lincoln called the Union to arms. The elder McGiffin volunteered again, this time elected lieutenant-colonel.[10] But after a period guarding railroads, Norton was invalided out with dysentery. In 1864 he left Pennsylvania and bought land in West Virginia, but six years later moved back to a farm north of Washington.[11]

National Guard riot service uniform

Philo was a frail, thin boy, a voracious reader, tall for his age but plagued by childhood illnesses. The youngest of five,[12] he was outshone by his rambunctious older brothers, Nathaniel, Tom, and Jim, and cosseted by his sister Sally.[13] He got good grades in the one-room school a mile and a half's walk away, and dreamed of going to sea; his schoolmates remembered the ships he carved in his desk with his pocketknife.[14] Once he and another boy decided they'd go West and fight Indians, with an old revolver and fifty cents in their pockets. His friend turned back at dusk, but Philo kept on until a farm couple persuaded him to go home the next day.[15]

When he exhausted the local school, Nort, as he was now called, enrolled at Washington and Jefferson College. He also joined the National Guard, just before Company H, Tenth Pennsylvania, was called out for a massive labor

Burning of roundhouse, 1877, from Harper's

strike in nearby Pittsburgh.

The Great Railroad Strike of 1877 was the response to layoffs, repeated wage cuts, and simultaneous dividend announcements by the B&O and Pennsylvania Railroads during a summer national unemployment reached 25%. When local militia, such as Philo's, sympathized with the strikers, Governor Hartranft called in units from Philadelphia. On July 21st the Philadelphians fired into a crowd, killing twenty, including a woman and three children. Thousands of furious workers and sympathizers broke into gun shops, arms factories, and armories,[16] and converged on the Pennsy yards.[17]

Part of Philo's company was cut off in a roundhouse and surrounded by the angry strikers. McGiffin volunteered to go for help. An engineer showed him how to start and stop one of the locomotives, then jumped off.[18] The seventeen-year-old drove the engine through the mob and got help.[19] Meanwhile the Philadelphia troops, surrounded by five thousand armed men, opened fire with Gatlings. The strikers poured kerosene on a trainload of coke, set it on fire, and rolled it in at the troops.[20] Twenty-five more

Destruction of Union Depot in Pittsburgh, 1877, from Harper's

died, including five soldiers, before the Philadelphians extricated themselves. After they left, the strikers looted and torched every railroad office, shop, and freight car in the city.[21]

The tall seventeen-year-old's first taste of combat had been against fellow Pennsylvanians. The day after he got back, Philo told his father he wanted to go to Annapolis.[22] The well-connected Norton had no trouble arranging this, and Philo was appointed to the United States Naval Academy on Sept 11, 1877.

In 1878, when Philo was sworn in as a cadet-midshipman, the Academy resembled nothing we know today. Flagg's wholesale reconstruction—the Mahan complex, Dahlgren, Macdonough, the Chapel, Bancroft—lay twenty years ahead.[23] The only buildings Philo would recognize in today's Yard are the brick guardhouses at Gate Three.[24]

Philo was called "McGiff" or "McGivv" by the cadets.[25] Admiral William Fletcher recalled many years later that his classmate had excelled in "Frenching," going over the Wall after taps and back without getting caught.[26] Philo was lively, well-liked and well-respected in an era when courage and honor were valued over grades, though he did well in his professional studies—when they interested him.[27]

In those days an abandoned mansion lay a mile upriver. One night Philo and his classmate A.J. Jones went over the Wall to see if the rumors of ghosts were true. Later their message was found scribbled on the walls: "This house is not haunted. P.N. McGiffin A. J. Jones."[28]

An incident one classmate thought illustrated "boldness and readiness of mind" happened when a Lieutenant Heald, an Officer of the Day "noted for his strictness and severe methods," found a bottle of whiskey during a night inspection of the cadet quarters. Heald returned to the Administration Building, where he was seen "lolling back in his chair, his feet on the table on which the bottle of whiskey rested in full view." Though the liquor wasn't McGiffin's, he decided to take action, and rang the fire bell. As Heald and his Marine orderly rushed out, Philo "entered the office by a window, seized the bottle, and escaped to his quarters. The purloiner was never discovered, although the cadets knew it was McGivv."[29] However, he'd damaged a hose cart while getting up through the window. The Commandant announced someone had to foot the repair bill, but the whole class volunteered to have twenty cents each deducted from their pay.[30]

The Academy had a large steam fire engine, several stationary steam-pumps, and frequent drills, with the cadets as the firemen. One drill was called away during dress parade. The "fire" was in the New Quarters. The cadets manned the pumps, and two streams were soon aimed, one at each end of the building. They crept toward the middle, and "...finally one crowd of nozzlemen, led by McGiffin, deliberately turned the hose on the other party. The return was made promptly, and all the disengaged cadets flocked to the fray." An officer who tried to intervene "nearly got drowned" but the fun continued until the steam was turned off.[31]

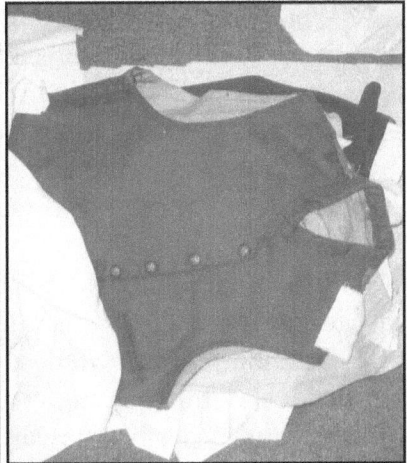

Philo as a midshipman *Philo's USNA blues*
Photo credit: James R. Norton

    A pyramid of cannon balls left from the War of 1812 was stacked on the upper floor of the Cadet Quarters. One summer night when McGivv couldn't sleep, he "decided that no one else should do so, and, one by one, rolled the cannon balls down the stairs. They tore away the banisters and bumped through the wooden steps and leaped off into the lower halls. For any one who might think of ascending to discover the motive power back of the bombardment they were extremely dangerous." But an officer caught him at last, and Philo was sentenced to *Santee*, the old guard-frigate where restrictees slept in hammocks 'tween-decks.[32]

    Cyrus Brady '83, who stood restriction with McGiffin for "some months and there got to know him very well," told about another restrictee who was "without doubt the most stupid man who ever attempted to go through the Naval Academy." (He spent four years as a plebe and never managed to get to third class.) Brady, Philo, and Cadet X were walking to recitation when their dull compadre asked McGiffin what the geometry lesson was. McGiffin told him it was about the three kinds of triangles. X asked what those were.

"Well," said McGiffin, "there are right-angled triangles like this," turning to the right and drawing imaginary lines in the air; "left-angled triangles like this," turning to the left and repeating the process; "and isosceles," which he described. X---- applied to me for confirmation—and received it, of course—and we tutored him all the way to the classroom...."

Unfortunately for McGiffin the officer in charge of the geometry section was a Civil War vet not noted for his mathematical knowledge. "Man the boards," he said, and the very first question, which he gave to X-----, was to name and describe the various forms of triangles. The cadet confidently rattled off the bum gouge his restriction buddies had fed him.

"Ahem!" said the officer, looking carefully from him to the book, "Mr. Chauvenet doesn't say anything about it, but the distinction you have drawn appears to be a very nice one. That will do, sir." So X---- made a very much better recitation and got a very much higher mark on that occasion than either McGiffin or I received!"[33]

Philo, who was both a good boxer and a skilled fencer, was asked to referee a grudge match. But the fight had to take place after supper, and McGiff was still restricted. He reported to Dr. Ruth with a "fearful looking tongue, white and ghastly." But the surgeon got the truth out of him: it was tooth powder. He listed McGiffin as off restriction due to illness, on his promise to report himself "cured" the next morning.

Once McGiffin was issued a new jumper. Pressed for time, he wrote his name on it in pencil. He was put on report for "not having name on jumper in sufficiently large letters." The next time he appeared, his name was printed in letters nearly a foot high, starting in the small of his back and going all the way around to the front.[34]

McGiffin's written assignments were professionally flawless, but he couldn't refrain from smarting off on paper, too, leading to more restriction. All this time on the old frigate led to the episode of the Cannons, Philo's best-known prank, retold in the Brigade in embroidered forms to this day. After listening to all the sea stories *Santee's* chief boatswain, Mike Morris,[35] had to tell, McGiffin talked the old salt out of six black powder charges to celebrate the Fourth with. (Probably saluting rather than full charges, which would have been hard to smuggle anywhere.) But Philo couldn't wait for Independence Day. "These he loaded into the six big guns captured in the Mexican War, which lay on the grass in the center of the Academy grounds, and at midnight on the eve of July 1st he fired a salute. It aroused the entire garrison, and for a week the empty window frames kept the glaziers busy."[36]

There's also a story that when a professor's house caught fire, McGiffin saved two children, and was commended by the Secretary of the Navy.[37] But Admiral Fletcher recalled no such fire, and there's no hard evidence of any commendation.

As Graduation neared McGiffin's class rank suffered, no doubt from all that time Frenching instead of sleeping. By the time it rolled around for the Class of '82, he stood 28th out of 36.[38] Still, many low-rankers have done brilliantly in the Fleet. But two months after graduation, disaster struck. Responding to officer overmanning in the moribund, shrinking Navy, Superintendent C. R. P. Rogers called for a reduction of class sizes and a cutback in the number of graduates getting commissions. This suggestion was adopted by Congress as the Act of August 5, 1882. Only the top 25% of each class would be commissioned, and only after two additional probationary years at sea.[39] The remaining 75% would be put out on Civvy Street with a year's pay in their pocket.

It was a threat likely to focus the mind of a young man who dreamed of glory.

The graduates bilged by the Act of '82 did not go gently. They hired lawyers, lobbied, organized, and wrote memorials. But the Senate rejected the reinstatement bill. The cause was lost.[40]

Meanwhile McGiffin had been putting in his two years at sea. He sailed with *Hartford*, Farragut's old flagship, calling at Funchal, Montevideo, Valparaiso, and Callao. This division officer tour faced him with gales, a yellow fever epidemic, faceoffs with drunken Chileans, and a stray round that just missed him during target practice.[41] He transferred to *Pensacola* at Callao for calls at Honolulu, Yokohama, Kobe, Hong Kong, Singapore, Batavia, Madagascar, and Cape Town before returning to Norfolk in May 1884. Philo came home with a trunk of exotic spears and daggers.[42] He went to Annapolis for his post-cruise examinations, with a recommendation from *Pensacola*'s CO, but didn't make the cut.[43] He strolled the Yard one last time, then headed for Ida Grove, Iowa, where his family lived now.

In Iowa McGiffin tried his hand at grain futures, then at politicking for James G. Blaine. Neither panned out. It's hard to say what turned his thoughts to China. Maybe it was the classmate's sister who'd been born there and praised the country to him.[44] But after reading a report of China's war with France over Tonkin (today's Vietnam[45]), he suddenly headed west. He stopped in Des Moines for accreditation as a correspondent with the *Iowa State Register*, then called on the Chinese consul in San Francisco. He stopped again in Honolulu to see his brother. Tom had been shunned by the family, possibly because of a liaison with a Hawaiian woman. But this didn't faze Philo,[46] who never showed any sign of the "Anglo-Saxon" racism then common.

Philo landed in Shanghai and took a small steamer north. *Waverly*, owned by the Chinese, with a Danish captain, under a British flag, was carrying contraband beneath a cargo of jute matting and manila paper. A blockader boarded her off the Yang-tse. Philo interpreted.

Governor Li Hung-Chang at Tientsin

He convinced the French he was a correspondent, and when the sailors unlading the holds onto the main deck got near the contraband, he persuaded the boarding officer to stop by warning he was endangering the vessel's stability.[47] *Waverly* was chased again by gunboats while approaching Tientsin (modern Tianjin).[48] She also struck a mine, but it was a dud.[49]

Unfortunately for McGiffin, peace was signed the day he reached Tientsin.[50] By then he was broke, and, as he put it in a letter home, a gone raccoon if nothing turned up. But as a last forlorn hope, he managed to attach himself to Waverly's captain, who was headed up to the palace to see the governor, Li Hung-Chang.[51]

Li was exactly the right man for Philo to see. He wasn't just a governor, he was the foreign minister too[52], and also charged with building a modern army and navy.[53] Dominated by the vain and venal Empress Dowager Tzu Hsi (Ci Xi), the Qing dynasty was tottering, but some officials were trying to modernize. Li held a literary degree, but was no pallid scholar. Ambitious and self-confident[54], he'd won face as a general during the Taiping Rebellion, personally leading his troops against rebel towns.[55] Since the 1860's, he'd led the "self-strengthening" movement, which tried to introduce Western technology in the form of arsenals, railroads, shipyards, machine factories, and military schools.[56]

Li had British and German instructors, and had tried before to recruit US officers.[57] Now 62,[58] he'd been dealing with Americans for twenty-four years, and liked them;

they seemed more honest to him than other white men and had no territorial ambitions.[59] Philo and the captain sipped tea with Li, then everyone lit up and sat back. A vice-consul interpreted. Philo wrote:

When it came my turn he asked: 'Why did you come to China?' I said: 'To enter the Chinese service for the war.' 'How do you expect to enter?' 'I expect you to give me a commission!' 'I have no place to offer you.' 'I think you have—I have come all the way from America to get it.' 'What would you like?' 'I would like to get the new torpedo-boat and go down the Yang-tse-Kiang to the (French) blockading squadron.' 'Will you do that?' 'Of course.'[60]

Impressed with McGiffin's aggressiveness, Li offered him the job, but recoiled when the tall, self-confident American told him he was only twenty-four. He needed experienced men to train his army and navy.[61] The doubtful governor-general sent him for what Philo called "a pretty stiff exam" in seamanship, gunnery, and math, by Li's naval secretary. He passed easily, and was told to report as a professor in seamanship and gunnery.[62]

The Tientsin East Arsenal[63] was enormous; the outer wall was four miles around. The northern naval college was inside, surrounded by a moat and another wall. But apparently not much of one; Philo wrote, "I thought to myself, if the cadet here is like to the thing I was at U.S.N.A., that won't keep him in." The Director commissioned him a lieutenant in the Imperial Chinese Navy, at a salary of $1800 in gold, plus expense account, house, rickshaw, and servants. He was also given command of that new torpedo boat until her captain arrived from England.[64]

McGiffin quickly found the Chinese cadets were not like those at Annapolis. Little else was, either. The torpedo boat was spotless, but when he tried to get underway he found no coal aboard, though it had been paid for many times over.[65] The cadets were physically soft and the dregs of the Imperial examination system. Worst of all, they feared fighting.

Over the next years Philo worked hard, teaching not only seamanship and gunnery as one of four foreign teachers,[66] but English composition,[67] fencing, and artillery and infantry tactics. He taught himself photography and learned Mandarin. He also campaigned to move the naval school to the coast.[68]

Philo gained Li's trust and was promoted. In 1886 Li sent him to Korea commanding a 120-man coastal surveying party. He caught malaria, recovered in Nagasaki, and visited the Japanese naval college there.[69] The next year he was sent to England with his trainees, whom he'd drilled in handshaking and polite conversation. They brought back *Ting Yuen* and *Chen Yuen*, two heavily armored 7600-ton pre-dreadnaught battleships, built in Germany for the Chinese,[70] that were the largest and most powerful warships in the Far East.[71]

On Philo's return, Li ordered him to build a new school at Weihaiwei. He battled larcenous contractors[72] for two years, varying this duty with supervision of survey training off the coast of Korea.[73] Phil opened the Peiyang Naval Academy[74] as its superintendent in 1890. He based its regulations on those at Annapolis, including the demerit system.[75]

Philo taught at Weihaiwei for four peaceful years. His Thanksgiving feasts, open to Americans and Chinese alike,[76] became famous from Port Arthur to Hong Kong.[77] He inspected the Northern Fleet and accompanied it as far afield as Vladivostok and Singapore.[78] He grew roses and played the flute in his walled garden, played with his little dog, Gyp, and experimented in his darkroom.[79] Promoted again, he had the blue orb of a third class mandarin and three gold stripes (in the form of a coiled dragon) embroidered on his dress blues.[80] All its buttons bore the dragon and anchor except the top one, which carried the eagle and anchor.[81] He looked forward to the leave he'd get in 1894, which he planned to spend in the States.

Meanwhile Japan was drawing up its plans against the tottering giant.

Philo with his dog in China.

In the summer of 1894, a rebellion sent both Chinese and Japanese troops rushing to Seoul. The Japanese got there first. They seized the Korean king, attacked a Chinese cruiser, and sank a transport loaded with troops.[82] A week later they declared war. McGiffin had packed for leave, but wrote his father it was a "point of honor" to stay with the men he'd trained.[83] He reported aboard *Chen Yuen*, one of the battleships he'd sailed back from Europe, of which he'd been appointed advisory captain or co-commander.[84]

McGiffin had struggled with graft and shoddiness in the Northern Fleet, too. In charge of ammunition, he'd managed to get a better grade of powder, but the shells were still defective.[85,86] Some were filled with black sand[87], others cement[88], and the supply was very short—there were only three explosive shells for the biggest guns, between the two battleships.[89]

The Peiyang (North Ocean) squadron was commanded by Admiral Ting Ju Chang, a former cavalry officer who didn't even pretend to understand naval tactics,[90] but was an old chum of Li's.[91] Ting's orders were to avoid the Japanese. His fleet consisted of fifteen ships, from ironclads down to torpedo boats. After joining up with more gunboats and torpedo boats, the fleet convoyed troopships to the mouth of the Yalu River. Escorted by the shallower-draft units, the convoy headed upriver while the deep-drafts anchored.[92]

Aboard *Chen Yuen*, McGiffin and his friend Captain Yang Yung Lin discarded woodwork and rigging, unshipped

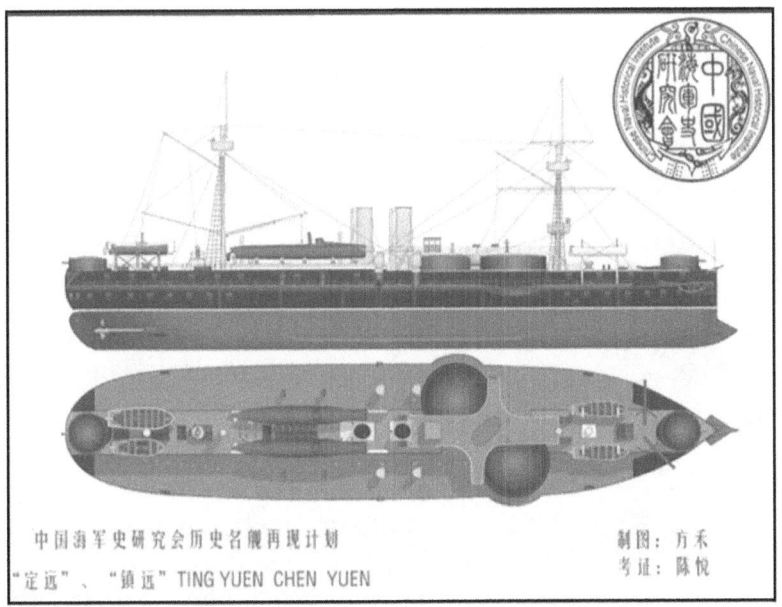

Layout of Ting Yuen *and* Chen Yuen

Ting Yuen *replica, 2005*
Photo by Peter Hunt, Hong Kong Society of Wargamers

Philo as Chinese Agent to England.
Photo credit: Dean McGiffin

glass, stacked sandbags and coalbags as splinter shielding, and painted her "invisible gray." McGiffin had hoses laid out for instant response if a fire started.

The cooks were making lunch on Sept 19, 1894, a fine day with a light breeze, when the lookouts reported coal-smoke on the horizon. The bugle sounded general quarters and the stokers began shoveling. The Japanese fleet, twelve ships in two squadrons, was approaching from the east, in line ahead, almost across the river's entrance.

Ting charged straight out at them in line abreast, the slower units lagging. On the bridge, McGiffin played waltzes on a music box to break the tension. After a time he put it away, took out his Book of Common Prayer, and led the Christians aboard in the Prayer to Be Used Before a Fight at Sea.[93]

The Chinese opened fire at five thousand meters, a sub-lieutenant estimating the range from the foretop with a sextant. The Japanese waited five minutes before returning fire.

Philo took some pictures before his Kodak was blown out of his hand. He retrieved it and took a couple more, then returned it to its case.

Two Chinese ships, one slightly damaged and the other not at all, turned tail and ran. The remaining eight steamed for the enemy at five knots. The Japanese traversed their front at twice that speed, then angled to starboard, turning Ting's right flank while firing heavily. Two old-fashioned cruisers on that wing burst into flames.

The Japanese flagship, *Matsushima*, doubled the right flank and led the rest of the squadron around the Chinese, cutting off any retreat. *Hiyei*, the slowest, broke the Chinese line. It took heavy damage passing between the ironclads, but made it though.

Philo was conning and fighting *Chen Yuen* now. As he wrote, "my Chinese colleague cheerfully gave me the brunt of it ... It was curious to see the effect of the first Japanese shell on the officers ... hardly one was seen after that."[94] His station was in the conning tower, but he went into the barbettes now and then to encourage the gunners.

Admiral Ting, 1894

The battle ground into chaos. Ting and his British advisor were knocked unconscious, decapitating the squadron.[95] The heaviest Japanese ships steamed around the largest Chinese, pouring in shells, ignoring the lighter vessels. But McGiffin maintained station even when surrounded. He feigned loss of steering control to lure the enemy flagship in, then hit *Matsushima* with a 12.5-inch shell that killed scores, started a fire, and forced Admiral Ito to shift his flag.[96] *Chih Yuen* charged the Japanese to ram, took a waterline hit and went down. Her captain made it to a floating oar, but was drowned by his dog. All this time what McGiffin called a "hurricane of projectiles from both heavy and light machine guns" and fragments from hundreds of 4.7 shells were causing heavy casualties on deck. Yet the Chinese enlisted stuck to their guns.

By three o'clock two more Chinese ships caught fire and went aground.[97] *Lai Yuen* caught fire, but stayed in action.

Chen Yuen's forecastle burst into flames too. None of her officers wanted to go, "and the men, of course, never like to go alone,"[98] so McGiffin, cursing, called for volunteers. He sent a lieutenant to stop the forward barbette from firing, and led the fire party forward.

He was bent over pulling up hose when he was wounded, but what really did him in was a muzzle blast from the 12.5s, which had not gotten the order to cease fire. He found himself on the deck, looking down the tube as the gunners reloaded. He rolled over an eight-foot drop[99] to avoid another blast, but landed on his face[100] and was wounded

Philo's Chinese dress blues;
detail of rank insignia and dragon and anchor button

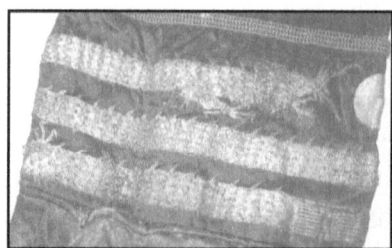
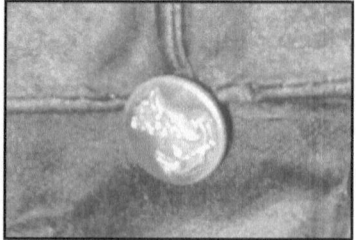

Shell holes in the superstructure of the Chen Yuen

again, more seriously, by another Japanese shell. He lay for a while bleeding, nearly blinded and only intermittently conscious; then was discovered and taken below. Captain Ling resumed the conn. Toward dusk the Japanese, fearing torpedo attack and low on ammunition,[101] withdrew. What was left of the Peiyang squadron limped into Port Arthur to lick its wounds.

Philo spent two weeks in hospital. When he checked out, though not nearly recovered, he found the damaged ships unrepaired and the wounded warehoused in filth and neglect. Infuriated, he managed to get hospitals set up for the men, but it was plain the fleet was finished.[102]

Analyzing the engagement afterward, McGiffin concluded the Japanese not only had a 50% numerical advantage after the Chinese desertions, but were superior in quick-firing high-velocity guns. His side's guns were of larger caliber but lower velocity, and a slower rate of fire. The Chinese had good crews and had fired more accurately, but the Japanese had outmaneuvered Ting and had better officers and shells that exploded.

Whatever the tactical lessons, the Chinese had not done well.[103] Philo's friend Captain Ling shot himself after the surrender of Weihaiwei and the remainder of the fleet, as did others.[104] Admiral Ting took poison. One captain who ran was beheaded.[105] The Japanese destroyed the Peiyang academy.[106]

The lost war cost China Formosa, Korea, the Pescadores, and a $300 million indemnity,[107] ended her naval ambitions for a hundred years, and may have cost her the last chance for a peaceful transition to a modern society. There was blame to go around,

Philo in bandages.
Photo credit: James R. Norton

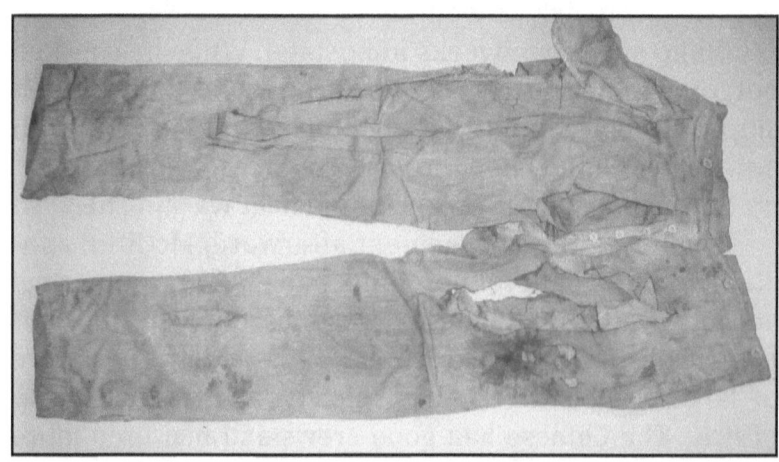

Philo's Yalu Trousers. Photo credit: Washington Historical Society

Chen Yuen *after her capture by the Japanese*

starting with the Empress Dowager, who'd done a massive renovation of the Summer Palace with the funds earmarked for the navy,[108] the crooked contractors and officials ashore, Li's own nepotism and skimming,[109] the incompetent Ting, and the cowardly line officers. But like most bureaucracies, the mandarinate pushed the onus downward till it reached those who couldn't fight back. (Of course, we only have the Western view; it would be interesting to research McGiffin and the battle at the First Imperial Archive in Beijing.)

This buck stopped with the foreign advisors.[110] Mandarins suggested to McGiffin that he take his life, as Ting had.[111] Instead he submitted his resignation. The commander of a British man of war got him through the blockade to a neutral port,[112] where he headed home via London.

McGiffin arrived in New York to a hero's welcome, and wealthier than when he'd left. But he was also ten years older, and his wounds weren't healing.

He got back to Washington in May of '96. The burly McGiffin of Weihaiwei was now gaunt almost to emaciation,

Marble boat at the Summer Palace

"Our Forces' Great Victory in the Battle of the Yellow Sea."
Japanese Illustration

and always in pain. He joked to friends: "I am still in the Chinese Navy; but I am not in good standing. You know it is customary for a naval officer, when he loses a battle to commit suicide; and they wanted me to follow the custom, but I declined with thanks."[113] A hometown friend recalled, "His abdomen seemed to be filled with pieces of steel . . . the skin was rough and ridged and hard. You could even see steel splinters sticking through the skin."

McGiffin seemed depressed, but set to work on an article for *Century Illustrated* on the battle.

McGiffin had always written well. The *Illinois Register* had published his travelogues. Behind his lucid prose lurked a deadpan humor like Mark Twain's. Journalism might have satisfied McGiffin's yen for adventure. It did for his contemporary, Winston Spencer Churchill, and for Philo's cousin and boyhood hunting partner Richard Harding Davis, who became a famous war correspondent, covered conflicts around the world, and published thirty-nine books. McGiffin had the contacts, the temperament, and the talent, but he was in increasing pain, still losing weight, and his sight was failing.

He accepted a commission to test a new solarometer on a cruise to England aboard SS *New York*.[114] When he returned, he lectured at the Naval War College on the handling of modern ships in battle.[115] A Colonel Pope offered him the managership of the Hartford Cycle Company in Connecticut.[116] But the pain worsened. His sister said later that "a bone was broken in his face when he threw himself to the deck to save his life from that gun, and there was another wound that never healed."[117] Several acquaintances saw him mine metal from his flesh with his penknife. After a year he went to New York for treatment.[118] He had an operation to drain an infection in the facial bone; this involved the surgeon boring a hole up into his head from inside his upper jaw, without anesthetic.[119]

McGiffin took a room at 49 West 33rd, hired a stenographer, and tried again to write.[120] He worked on pieces for *Century* and *The Strand*, and corresponded with Fred Jane of the *Naval Annual* about exaggerated accounts of his role at Yalu that sensationalist reporters had published. Cuban revolutionaries sought him out, hoping he might fight for them; he turned them down courteously.[121] He started a memoir, but gradually slid into a lonely, pain-ridden funk.

At last he reached out to an informal network of USNA alumni. When Park Benjamin ('67) visited, Benjamin described a man with "mutilated body and shattered mind," pacing night and day as he re-enacted the Yalu fight, cursing Chinese corruption, treachery, and cowardice. "He got his revolvers and prepared to repel imaginary assailants."[122] Someone wrote his brother. Jim came to New York, and with the assistance of an alum who was a director,[123] checked Philo into New York University[124] Post-Graduate Hospital at 20th and Second (now Peter's Field Park).[125]

The director wrote to an enquiring midshipman in 1929 that McGiffin had been injecting himself with morphine for a long time, and "the habit had become fixed upon him." When McGiffin was admitted, his doctors feared he might become suicidal as the "opium" was removed, and took away two guns he had with him.[126] (Both opium and morphine were readily available in the 1890's, and legal until 1914.[127])

After several weeks his health began to improve.[128] McGiffin befriended another patient, a woman, and they exchanged books and notes.[129] The doctors got the infections under control. But whatever preyed on him—fear of blindness, depression, unbearable pain, unwillingness to submit to more surgery, morphine withdrawal, or his internalization of Chinese mores—won out.

Around 2 A.M. on February 11, he asked for a tin dispatch box containing his manuscripts. Since they'd already confiscated two guns from him, no one bothered to search it. Examining the papers, Philo feigned falling into a doze. The attendant turned down the light and tiptoed out for a snack.[130]

No one heard the shot. But when the nurse returned, he found McGiffin slumped, blood running from his right temple. A smoking Smith & Wesson lay on the nightstand, a Navy Colt on the floor.[131] Also on the nightstand was a page from his notebook. "Memo—look alive that the bed is not

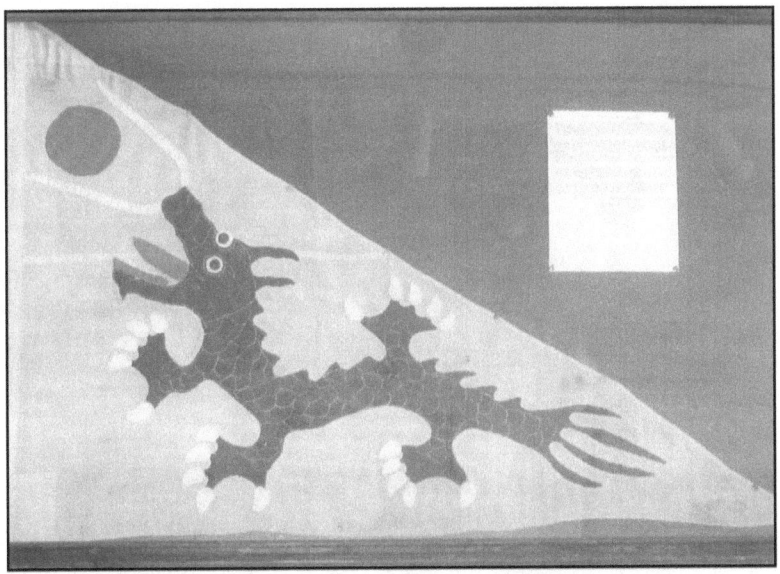

Chen Yuen's *Naval Ensign*

set fire by my shot. My compliments and adieus to all. I regret that my destination must remain to you unknown—but you may guess. With apologies to Miss Phelps for the row—it is the way that all guns have. Au revoir." On the other side was, "Please give my good-bye to Mr. Gorham. My compliments to Mr. Bennett. Incidentally I may note that though I effect my exit it is not the hegira anticipated by the learned staff."[132] (Gorham and Bennett were his nurses, Miss Phelps the matron.[133])

In Washington, Pennsylvania, McGiffin's body lay almost in state at his father's house. It weighed only ninety pounds.[134] His family dressed him in his Imperial uniform and draped the casket with *Chen Yuen*'s flag.[135]

Philo McGiffin is buried in Washington Cemetery, under a stone decorated with the American and Dragon flags. He wrote his own epitaph: "A broken and a contrite heart, O God, thou wilt not despise."[136] His memoir, the book he'd worked on for that last lonely year, was stolen while he was in the hospital, and never recovered.[137]

Philo's marker in Washington, PA

Philo's gravestone

Years later Norton McGiffin, Philo's nephew, was Marc Mitscher's roommate at USNA for a year before developing eye trouble.[138] Norton's son Donald served in the Navy in World War Two.[139]

Probably no one will ever improve on Park Benjamin's final tribute to his friend:

"Had he lived a couple of centuries ago there would have been room for his sword with many a gallant White Company; or his colors might have proudly flown over many an adventurous and bloodstained deck. When he came to realize that those days are gone, the world seemed to have no place for him; and so, being very tired and very sore both in body and in mind, he touched the merciful trigger and, with a debonair apology for his unceremonious exit, finished for himself the work which the Japanese had failed to complete."[140]

6

RICHMOND P. HOBSON, CLASS OF 1889: THE MOST KISSED MAN IN AMERICA

It sounds like a Trivia question. What do the battle of Santiago, Prohibition, prison overcrowding, and Hershey's Kisses have in common?

The answer is—Richmond Pearson Hobson, USNA Class of 1889.

His story is full of twists and turns, bravery, occasional silliness, and a hero many found all too gallant for their tastes. It begins in Greensboro, Alabama, in 1870.

Richmond Pearson Hobson was born on the western edge of town. Magnolia Grove still stands today, a temple-style Greek Revival home built around 1840 by a wealthy planter and slaveowner named Isaac Croom.[1] Croom's niece, Sallie Pearson, had lived there with his widow Sarah after Croom's death in 1863.

In 1867 Sallie married James M. Hobson. A University of North Carolina graduate, Hobson had enlisted in the Second North Carolina Infantry, Company E, on the outbreak of the Civil War. He was wounded at the battles of Malvern Hill, Chancellorsville, and Spotsylvania. Captured at Spotsylvania, he spent the rest of the war as a prisoner.[2] After Appomattox, he studied law under Sallie's father. In 1879, they bought Magnolia Grove, though Sarah continued to live there as well, and established their family.

Hobson's boyhood sounds very much like Tom Sawyer's, only more privileged, as befitted the scion of two distinguished families in a culture and time that placed great emphasis on lineage. According to his mother, he could "run the greatest distance, swim better and longer than any boy in town." Like most lads of the time, he spent a great deal of his time in the woods and fields.[3] He and his friends fished with live wasps for bait, flew kites, hunted for squirrels with an old muzzle-loader, went barefoot all summer—"our parents thought it made us hardier"—and played at knights with sharpened sticks for lances.[4] He also distinguished himself in memorizing Scripture and poetry. Photographs show him as a handsome, serious-faced young lad. A historian who knew him then described him as ". . . grave faced. His manner was stiff and formal; his conversation almost comically stilted."[5]

During a visit to New Orleans, Hobson was strolling Canal Street with his school friends when they heard a man calling, "Visit the man-of-war for twenty-five cents!" They jumped on a tug and were taken aboard *Tennessee* for a tour. "Rich" soon began talking of joining the Navy. (Hobson himself, in his autobiographical novel *Buck Jones at Annapolis*, seemed to say that as his older brother had decided for West Point, he felt it his duty to go to Annapolis.[6]) Be that as it may, he entered Southern University at the age of 12, attended for three years, and won a declamation prize.[7] In 1884 he was appointed to the Academy by competitive appointment.[8]

Hobson arrived at USNA three months before his fifteenth birthday, the youngest member of his class. Years later, he said in Congress that he had "run the gamut of hazing at Annapolis in perhaps its severest form."[9] A mid who abstained from alcohol, prayed each evening on his knees, and read from the Bible was a natural target for upperclassmen, and he was promptly dubbed "Parson."[10] Unlike Philo McGiffin, the idea of ridiculing the Administration, playing a practical joke, or being anything other than the straightest arrow in the quiver, seems never to have occurred to Richmond. Still, once plebe year was over, he found the academic load no great challenge, given his intelligence and voracious study habits.

What tripped him up once again were what some might call his unbending moral rectitude; others might have had a less friendly name for it at the time. Hobson ended his first day as Officer of the Deck having papped a full two large pages of his own classmates for such offenses as smoking, being late to formation, improper uniforms, etc. In those days, putting one's own classmates on report did not earn style points, and Hobson was voted into Coventry that same day.[11]

"Coventry" is one of those practices that has faded unlamented from living memory. The object of shunning was not spoken to or eaten with, and even had to swim alone. Hobson remained in that status for two years.[12] "He went about in silence, treated with scorn and contempt."[13]

The aim was to force a resignation by social ostracism, but his classmates had not taken Hobson's measure. Maintaining a rigid schedule of study, reading, gym, swimming, and horseback riding, he seemed not just to endure, but even flourish. By that summer another mid, Sumner Kittelle (later Rear Admiral Kittelle, who gave DesRon 11 orders that resulted in a famous mass grounding incident), had joined Hobson in Coventry, and one by one others came too, making the ostracism moot. Meanwhile, Hobson became president of the Academy YMCA and Cadet Battalion Commander (today's Brigade Commander) during his first class year. He also graduated first in his class.[14] By then, apparently, those who had cast him out had invited him back among them, but Hobson said he'd gotten along without them, and saw no reason why he could not continue to.[15]

Years later another classmate wrote: "He was always a crusader, always honest, always fearless, and because he was in a way different, he was not appreciated at that time by all of his classmates, to whom mediocrity and personal property meant more than honor."[16] Perhaps because of Coventry, though, Hobson chose, not Navy Line, but the Constructor Corps. He would be a staff officer, not a line officer. And it is difficult not to read that as at least some bitterness in a man rejected, not for any shortcomings, but for acting in a way he believed was morally right, even if misunderstood by his peers.

Hobson at first appeared destined for the highest distinction in the staff corps. After a year's cruise aboard USS *Chicago*, a 4500-ton protected cruiser, he was selected for postgraduate school—a rare honor then. He was the first American to graduate with honors from the Ecole d'Application du Genie Maritime. Supervisory work at various shipyards followed on his return.

However, this promise was marred when the Sino-Japanese War broke out (1894–5) and Hobson decided he wanted to observe it. He used the old-boys' network to get permission from Secretary of the Navy and fellow Alabaman Hilary A. Herbert. But when his more immediate supervisor at the Bureau of Navigation, Francis M. Ramsey, got the request, he disapproved it. (It was Ramsay who had tried to send Alfred Thayer Mahan '59 off to sea from the Naval War College the year before because "It was not the business of a naval officer to write books."[17]) Here the record gets murky, but apparently Hobson finally did make it to Asia somehow, and toured and reported on Japanese, Chinese, and European naval bases in the area.[18] He was unpleasantly impressed with what was called at the time the "Yellow Peril."[19] However, "His readiness to go straight to the top was making him enemies within the service."[20]

Hobson recommended that a three-year, postgraduate course in naval construction be started at USNA.

This was approved and he was selected to run it.²¹ This was the first postgraduate program at the Academy. At the same time, though, some castings he'd approved during the construction of *Kearsarge* at Norfolk failed, and he was formally reprimanded. Rather than sucking it up and soldiering on, he responded with a dramatic forty-seven-page statement addressed over his superior's head to Assistant Secretary of the Navy Theodore Roosevelt. Roosevelt upheld Hobson's senior and Richmond had to undergo a humiliating exam on machine tool practice.²²

Frustrated personally, and laboring now under a perception within the Service that he was both something of a prig and moreover, one who didn't mind going over his seniors' heads to the civilian leadership, Hobson may have seen the war with Spain which now loomed as an opportunity to redeem himself.

Hobson was aboard *New York*, Rear Admiral William Sampson's flagship, with his postgraduate students when war was declared in April of 1898.²³ He attached himself, or was attached, to Sampson's staff as Admiral Pasqual Cervera's seven-ship Cadiz squadron transited from Spain to defend Cuba against the expected invasion. After various countermarches and near-misses, by late May Cervera was reported tucked into Santiago Harbor, with Sampson's North Atlantic Squadron on a breakneck steam down from Key West.²⁴ The harbor entrance was narrow and so heavily defended with forts and mines a fleet could not expect to go in without heavy losses. Instead, Sampson asked his naval constructor to prepare a plan to bottle up the Spanish inside.²⁵

The best account of Hobson's epic exploit is in his own book, *The Sinking of The* "Merrimac." Hobson's prose is lean and the narrative is tensely told, with hand-drawn sketches of the channel and harbor, cross-sections of the ships involved, and detailed drawings and explications.

Time was short. The right combination of tide and moon would occur soon, and Sampson wanted Cervera bottled up quickly. By the time the Squadron reached Santiago, Hobson had a "scheme" to scuttle *Merrimac*, a converted collier, athwart the 350-foot-wide channel. Since any approaching ship would come under heavy fire from both the Spanish forts and fleet, he planned to charge in at full speed, with the tide behind him, and deploy anchors to brake and then swing athwart the channel. Finally, he would blow the bottom out with "torpedos" *New York's* gunner's mates were improvising from eight-inch powder charges. He would do this with a minimal, picked crew who, if they survived, would escape by small boat. And he would command the mission.

Arriving off Santiago on June 1, Hobson selected volunteers—mainly on the basis of their looks, to judge by his account—and took a steam-launch in so close to shore to reconnoiter that he drew rifle-fire. Returning, he boarded *Merrimac* to give her captain the news.

Hobson had only one night to make complicated preparations. Assisted by Robert Crank, Class of 1892, *Merrimac*'s engineer, and the crew from *New York*, he rigged the collier's anchors and deployed his jury-rigged powder charges over the side, to blow in the plates beneath the waterline. He sent around the fleet for electrical detonating boxes, but only batteries were available to fire the electric primers. Meanwhile he fended off repeated requests to accompany him on what was rapidly starting to look like a suicide mission. *Merrimac*'s captain, James Miller USNA '67, also felt he had to go, until told directly by Sampson he would not. Confusion reigned as the offgoing crew stripped the ship, as men helped themselves to souvenirs, and aged winches proved incapable of handling the heavy chain cables. A launch fouled its screw on the torpedo lines, further adding to the delay. Finally, as *Merrimac* built up speed for the run in, the escape lifeboat capsized and broke adrift. Hobson determined to go in anyway. But daybreak was now too close. Admiral Sampson signaled recall. Hobson contemplated disobeying the signal, but at last, reluctantly, turned back.

They spent the next day waiting around, tense and unable to sleep. There was little food and not even any coffee, a lack which affected Hobson, a teetotaler and nonsmoker who did like his joe. He found more batteries and got another lifeboat and a pontoon-type "catamaran" aboard to replace the one that had been lost.

After midrats the next night, at 1:30 AM, June 3, Hobson took them in again. A launch from *New York* trailed them in the darkness, during which they passed two boats manned by reporters.

Merrimac entered the channel just after moonrise, at her nine-knot top speed. The seven-man crew lay full length on the deck in underwear and revolver-belts, grasping the signal lanyards leading from Hobson, on the bridge, and the firing wires to the torpedoes. Four hundred yards off the channel entrance they encountered a picket boat, which immediately began popping away at their exposed rudder with its quick-firing gun.

The forts, shore batteries, and even the Mauser-armed Spanish troops quickly joined in. The collier was perfectly visible in the moonlight and the fire was heavy and accurate. Hobson described the noise: "The striking projectiles and flying fragments produced a grinding sound, with a fine ring in it of steel on steel."[26] Nevertheless, he held course until he felt she would make it into the channel under momentum alone, then ordered the black gang to smash the intakes, abandon the engine room, and take their positions at their assigned charges.

Target of hundreds of guns, *Merrimac* staggered into the narrow channel. She nearly hit Morro castle but swerved away. Hobson, on the bridge, discovered either the rudder or the steering-chains had been shot away just when most needed. Still, they were in the channel, being swept inward with the current. He coolly dropped the bow anchor, then ordered the torpedoes fired with three yanks on his signal lines.

Only two went off. Wires had been shot through and batteries shattered by the gunfire, which grew even heavier as *Reina Mercedes,* a cruiser anchored as channel guard, joined in. The chain of the bow anchor parted. The stern anchor didn't hold either. *Merrimac* staggered on, then lifted to an underwater explosion. A Spanish mine had gone off, one of at least ten command-detonated from shore as what the defenders took for an attacking cruiser zigzagged into the harbor.

Yet even that didn't sink her, to Hobson's rage. If only Sampson had let him bring along the hundreds of pounds of guncotton he wanted, as a last-ditch expedient! They grounded, then broke free and swept onward. He started forward to raise the Stars and Stripes, and was only talked out of it by the crew, huddled under the bulwarks, who felt it would draw fire directly to them.

The hulk was now under intense fire from three sides, and being thoroughly riddled. Though *Merrimac* was unarmed, the Spanish lost fourteen killed and thirty-seven wounded from their own crossfire that night. The men hugged the deck as more mines went off. Hobson watched the channel shores go by, seething. She was leaving the narrows. Why would she not sink?

At last two Whitehead torpedoes from the Spanish cruiser tore into her. Just as she emerged into the wider inner harbor, *Merrimac* swung sideways and went down. Hobson and her men were washed overboard, sucked down, vomited back up. Battered by debris, covered in coal-powder, they swam to the bobbing catamaran as the firing slowly subsided.

Hobson and all his crew got away alive. They clung to the catamaran, steadily losing body heat, until daybreak, remaining quiet as rowboats with lanterns searched the debris. Hobson had appointed a rendezvous in a sea-cave beneath the eastern cliffs, and may have had some idea of going ashore there, then making his way back to sea. Shortly after dawn, however, a steam launch headed for the catamaran. Hobson finally called out and was immediately covered by several seamen with rifles. They didn't fire, though, and he was able to surrender to an oldish man he was later to discover was Admiral Cervera himself. To him Hobson turned over his revolver and life-preserver. He was a Spanish prisoner.

Hobson and his men, only one of whom was even slightly wounded, remained in enemy hands for the remainder of the siege, while a U.S. Army force landed under General Shafter. His account emphasizes the courtesy and honor the Spanish showed him, but the circumstances of his captivity made strange reading to a modern eye. They included an open cell door, three-course meals served by an enlisted waiter, notes sent back and forth to Sampson via truce boat, telegrams received and sent, and courteous interviews during which his interrogators begged his pardon before asking questions. Meanwhile *Merrimac* continued to partially block the channel.

Despite the barrier, the Spanish sortied on July 3, and were promptly destroyed by the guns of the waiting Americans.[27] Shortly thereafter Hobson and his men were exchanged. He passed into the American lines cheered by the whole army, a scene reported by Stephen Crane and Richard Harding Davis, and a letter of congratulation from the SecNav awaited him aboard the flagship; but this was only the faintest foretaste of what was to come.[28]

Hobson may not have succeeded in blocking the channel, but his exploits made him not just a hero, but something entirely new. Here, at the dawn of the era of popular journalism, Hobson may truly have been the first American celebrity. His extraordinary handsomeness, his spotless personal life, the fact he was a Southerner—the first such to become a war hero since Appomattox[29]—his undoubted courage, the fact that he was single, even the touching story of his courteous rescue combined in a "perfect storm" to make him a national phenomenon from the moment he set foot back in the States. After a stopover in Washington, the Navy sent him across the country on a speaking tour, starting in New York. Hobson addressed a massive crowd at the Metropolitan Opera, and riveted them all. On his way out, a woman tried to kiss him, but failed. However, Hobson did meet another "Southern girl," who was later to become his wife.[30] But it was the next day he received "the kiss heard round the world."[31]

Overwhelmed by barrels of love letters and interview requests, Hobson tried to escape to the home of a family friend. However, that evening he was the guest of honor at a reception at the Long Beach Hotel. After accepting a presentation sword, he was ambushed in the reception line by a Miss Emma Arnold of St. Louis and inveigled into giving her a kiss.

The press had a story and they ran with it. Both New York and St. Louis headlined the story, and it was picked up nationwide. A waltz was composed in Hobson's honor, and a candy manufacturer put out "Hobson's Kisses," apparently a toffee confection—the chocolate version would come later, and continue in production to the present day.

A short lull, then activity picked up again when Hobson reached Chicago, kissed two of his cousins, and then went on to kiss the other women in line—163 of them, to be exact.[32] Kansas City was worse (or perhaps, better), with the score between 417 and 419. In Denver, the platform nearly collapsed beneath a thousand struggling women, and the newspapers had a field day both applauding, then questioning whether all this was in good taste. Hobson was now famous not just in the U.S., but throughout Europe as well.

But something had been lost, too. *The New York Times* observed, "There is something distinctly tragical to the fact that the name of Hobson now brings to the minds of a considerable part of this country's inhabitants, not the recollection of a heroic act rarely paralleled in military history, but the image of a fatuous youth submitting to or seeking the cheap caresses of a lot of foolish and vulgar women."[33]

Hobson was advanced ten numbers in rank for his exploit, becoming a captain at the age of twenty-nine, but making his position in the Staff Corps uncomfortable. His application to transfer to the Line was turned down and he was assigned as Inspector of Spanish Wrecks.[34] Over years in Cuba and the Far East he raised several wrecks, including *Reina Mercedes*, which had sunk him. She became the Naval Academy station ship in 1912. Moored at the seawall, she served as the "brig" ship until 1940, when restriction to Bancroft became the accepted disciplinary measure. She then served as an enlisted berthing barge, headquarters for the Academy's sailing activities, and harbor control center until decommissioned and sold for scrap in 1957.[35]

By 1903 Hobson was convinced he was going blind and that his career was at a dead end. He applied for medical retirement but was refused. He resigned and immediately went on the national lecture circuit across the country.[36] He used his oratorical talent and celebrity status to push the need for a large Navy to meet a rising threat from Russia and Japan. Interestingly, he saw Germany as America's natural ally against England and Japan, who were at that time treaty allies.[37] In 1904 he disputed the Democratic nomination in his congressional district in Alabama but lost to the incumbent. He ran again in 1906, emphasizing the necessity to fight monopolies and regulate the railroads. This time he won, defeating one of the most powerful men in Congress and causing speculation he might be a vice presidential or even presidential possibility for 1908.[38]

Hobson's congressional career extended from 1906 to 1914, a time of massive change. Unlike most Southern Democrats, he aligned himself with the Progressives. He attacked big business and voted for government regulation and against the protective tariff. He pushed naval preparedness against Japan from his seat on the House Naval Affairs Committee, in one case—at the Democratic convention in 1908—so hard he seriously angered Theodore Roosevelt.[39] Hobson introduced a resolution to abolish the Electoral College and directly elect the President, and was instrumental in removing the election of senators from state legislatures.[40] He was an ardent supporter of women's suffrage, and one of the leaders of the 1913 Washington Suffrage Parade that ended in a near riot.[41] He spoke widely in favor of world government and a universal peace compact.[42] He even rose in Congress to defend the "Brownsville Regiment," Negro troops accused of murder in Texas, saying: "Whether the heavens fall or the earth melt away, while we live let us be just." He was the only Southern Democrat to vote for a fair trial, and it hurt his chances for re-election.

More and more, though, Hobson became involved with his next great crusade, for which he became the principal advocate in Congress. He campaigned against alcohol in thousands of speeches, introduced the first Prohibition amendment in 1911 (it failed), and later wrote a polemical book, *Alcohol and the Human Race*.[43]

Alcohol is not agreeable reading. The facts he cites seem reasonable, but his philosophizing on them is turgid and overwrought, full of questionable generalities, such as that without alcohol, war will become a thing of the past.[43] A sample of his prose: "As the chief fundamental cause of immorality, alcohol is the mother of venereal disease. Mother and whelp are continually stalking through the land and, not in the slums alone but in tens of thousands of good homes, are tearing to pieces with foul fangs the holy weft of which nature weaves new lives."[45]

Once out of Congress, Hobson transformed himself into a professional temperance crusader. His stock speech was called "Alcohol, the Great Destroyer" and it rested on the idea that "races" succeeded in the struggle for existence until corrupted from within by alcohol. Alcohol damaged the "top part of the brain" and the reproductive organs. For America's bright future to be realized, the "great destroyer" must itself be destroyed in a great war.[46] This was a tremendously popular message and Hobson, the same brilliant and dogged organizer who had helped Sampson strip the Atlantic Squadron of every bit of flammable wood as it raced south to its rendezvous with fate,[47] carried it across the country in thousands of speeches and millions of pamphlets until Prohibition was finally enacted in 1920.

At which time he faced a new problem. He had received a large income from the Anti-Saloon League for his speaking,[48] but he'd worked himself out of a job. Hobson was never wealthy and had to support not only Grizelda and their three children, but also several members of his extended family and Magnolia Grove as well.[49] This is not to say he did not believe in his crusades, only that he also had to make them pay. His next target, accordingly, became narcotics.

Here again Hobson made a tremendous impact both as a speaker and as an organizer. He incorporated the International Narcotic Education Association in 1923 and was immediately elected president.[50] Hobson wrote "The Perils of Narcotics" and lobbied Congress to send fifty million copies of his pamphlet to children across America. His article "One Million Americans Victims of Drug Habit" was headlined across the country.[51] He worked with both parties to incorporate anti-narcotics planks in their platforms and was again boomed for president at the 1924 Democratic convention. He published *Narcotic Peril* (1925); *Modern Pirates—Exterminate Them* (1927); and *Drug Addiction, a Malignant Racial Cancer* (1933). It's probably fair to say that the oft-remarked difference between American attitudes toward opiates and European policies is that Europe never had Richmond P. Hobson. Continuing to speak widely, he also organized and chaired world conferences on narcotics and tried to stop use of cocaine by the makers of Coca Cola.[52] Progress was slow, but eventually the League of Nations ratified the Opium Convention in 1933 and invited Hobson to address its final conference in recognition of his efforts against the trade.[53]

For the remainder of his life, Hobson fought to achieve ratification of the Convention around the world and to strengthen U.S. narcotics laws. Future years would not see narcotics "conquered," Prohibition was repealed in 1933, and the Depression brought Hobson desperate money troubles, but at least a system of international control was established. In 1933 Congress promoted him to Rear Admiral, and awarded him a pension and the Medal of Honor. Hobson formed new associations to promote industrial safety and, once again, to build up the Navy against Japan.

Hobson fell in harness, collapsing and dying in the street in 1937 as he walked to his office in New York. He left an estate valued at less than $5000.[54] Rear Admiral Hobson is buried in Arlington. USS *Hobson*, DD-464, was commissioned in 1942, served in the Atlantic and Pacific (ironically, in the war with Japan Hobson had long predicted), and served until lost in 1952 in a collision with USS *Wasp* off the Azores.[55]

There's much to dislike about Richmond P. Hobson. He scorned accommodating vested interests, flattering his superiors, or getting along with his peers. To many he came across as holier-than-thou, stuffy, or lofty. He was prone

Hobson's Fearless brand cigars

to exaggeration and not overly concerned about factual accuracy as he pushed his various crusades.

Yet, on balance, there's more to admire. He was undoubtedly brave. He defied not just Spanish guns, but his own era's racism and misogyny. He was truly dedicated to the improvement and elevation of America and mankind as he saw it.

Even on Valentine's Day, when he is justly memorialized, Richmond P. Hobson deserves to be remembered as something more than the most kissed man in American history.

WENDELL NEVILLE, CLASS OF 1890: "RETREAT, HELL!"

"An indomitable will, a sense of humor and conspicuous courage were joined with personal magnetism to make him a soldier of heroic mold." —The Washington Post

"Retreat, Hell! We just got here!" —Wendell Neville, 1918

There seem to be rather fewer Marine Corps heroes from our august institution than those who wore Navy blue. Such iconic Marines as Smedley Butler, Chesty Puller, Pete Ellis, Aquilla Dyess, and Holland Smith rose from the enlisted ranks. (In part this was due to statute; for a long time, officer vacancies were filled first by breveting from the ranks, and only after that from Academy grads.)[1] We do have some famous names, though. And for sheer gallantry over an extended period, one figure stands out. Wendell Cushing Neville, Class of 1890, definitely logged some fighting time over his thirty-eight year career. Guantanamo Bay, the Boxer Rebellion, Veracruz, the Philippine Insurrection, the Banana Wars, Belleau Wood—the names recall a whole era of struggle and triumph for the Corps.

"Wen" Neville was born in Portsmouth, Virginia, a shipbuilding town, on May 12, 1870. Unlike many Academy nominees then, he didn't come from monied circumstances; his father was a ship's carpenter. He had five brothers and a sister, and was raised Methodist. Neville's father died when he was thirteen, but an older brother paid for his schooling. He graduated from Norfolk Academy in 1885 and received an appointment to the Naval Academy from the Second Congressional District.[2] In later years, he said that he got the appointment because no one else wanted it, but this seems doubtful; in a traditionally Navy town like Norfolk there must have been many more applicants than one carpenter's son.

Neville arrived at USNA at the age of sixteen. He was ranked 29th out of 53 at the end of his youngster year, and 32nd of 43 his second class year. His demerits records reveal nothing out of the ordinary; his best grades were in algebra and geometry, his worst in English and strength of materials.[3]

"Not a particularly good scholar,"[4] Neville graduated number 22 of 31 in the Class of 1890.[5] He then served two years at sea as a passed midshipman, in *Kearsarge* and *Newark*, before returning for service selection. Unfortunately, a mediocre performance on the post-cruise exams gave him only two choices: civilian life, or the Marine Corps. (He told his daughter years later, however, that the reason he chose the Corps was his dislike for deck watches at sea.) In July of 1892 he and four other classmates were commissioned as second lieutenants.[6]

After three years aboard cruiser *Cincinnati*, battleship *Texas*, and cruiser *Raleigh*,[7] Neville reported to the Marine Barracks in Washington as the Guard Officer. There he met and promptly married (January, 1898) the daughter of the Commandant of the Navy Yard, Frances Adelphia Howell. Their only child, a daughter, also named Frances, was born the next year.[8]

As the Spanish War began, the Commandant ordered Lt. Col. W. R. Huntington to form a Marine expeditionary

Neville in command of USS Texas Marine Guard, 1896

battalion in New York.[9] Neville reported there, and the ad-hoc battalion was landed at Guantanamo Bay on June 10 to secure a harbor Admiral Sampson could coal at during his blockade of the Spanish fleet at Santiago.[10] Huntington first sent the young lieutenant to take charge of a crossroads dominating the valley.[11] Then, as Spanish sniping and harassing fire went on day after day, on 20 June he ordered Neville and two companies of Marines, plus fifty to seventy Cuban guerrilleros, to secure Cuzco Well, the enemy's watering point.[12] Shells began falling on the Marines from USS *Dolphin*, which had been assigned as their fire support, but a heroic sergeant wigwagged a "cease fire" signal. The Marine assault drove off roughly five hundred Spanish of the Sixth Barcelona Regiment at the cost of six KIA and sixteen wounded, and ended enemy resistance at Guantanamo.[13,14] Neville was mentioned in dispatches and breveted to captain.[15] (Brevet promotions were popular because at this time Marine officers were not eligible for the Medal of Honor, and no other award for gallantry existed until 1905.)[16]

Neville returned to New York and served briefly as officer in charge of the Marine Corps Recruiting Station there.[17] But a few months later, he was again hastily ordered off, this time to China.[18]

The China Relief Expedition was mounted to rescue four thousand diplomats, women, children, and Chinese Christians trapped in Peking by antiforeign rebels (The I Ho Ch'uan, or "Society of the Righteous Fists[19]"). In some ways, it was the first modern multinational task force.[20] As China rose against all foreigners whatsoever, Neville's Marines marched through intense heat from Tientsin to Peking. Four battles and grueling forced marches led to the eventual breakthrough on August 14; after helping to (unfortunately) loot and burn parts of the Forbidden City, the Marines were withdrawn to Cavite in September.[21,22] Not only was Neville again commended for gallantry, he made acquaintances which would be significant in his future, including Smedley Butler, William Biddle, Benamin Fuller, and a heroic young civilian named Herbert Hoover.[23]

The newly liberated Filipinos were disinclined to accept U.S. control in place of the independence they felt they'd been promised. Aguinaldo revolted in February 1899.[24] The Insurrection became a vicious war, with over ten times the U.S. casualties as the Spanish-American war itself.[25] But eventually Arthur MacArthur and Frederick Funston captured Aguinaldo and tamped down the revolt. Neville served as the military governor of Basilan Province from mid-1901 to late 1902. Information is scarce about what he did there, but apparently Basilan, interestingly enough a Moro region, did not resist administration and cooperated against Aguinaldo's revolutionaries.[26]

Neville, now nicknamed "Buck", emerged from China and the Philippines as a seasoned tropical campaigner.[27]

Tall and imposing, he could present a gruff appearance at first meeting, but also had a reputation as the most cheerful man in the Corps.[28] "Nobody could be downhearted when Neville was part of the company," one acquaintance would later say.[29]

He served briefly at Headquarters, Marine Corps, and umpired at the War College. Promoted to major in 1904, he commanded the Rhode Island Barracks, then commanded detachments aboard battleships *Maine* and *Connecticut*. He returned to Cuba again in 1906, when Roosevelt ordered Havana occupied.[30] He commanded the Marine Barracks in Washington, took a battalion to Panama and Nicaragua to join Maj. Smedley Butler and Col. Joe Pendleton (USNA '84) for short actions in 1909 and 1910, and established the Marine Barracks at Pearl Harbor.[31]

Also of interest during this period was his opposition to President Theodore Roosevelt's executive order removing Marines from ships as a preliminary to transferring the whole Corps to the Army. Neville and several other prominent officers organized to defeat this proposal, a union that led to the founding of the Marine Corps Association.

In 1909 Neville, described as a "thin-faced marine with close-cropped light brown hair,"[32] was ordered to sit in court-martial at the Naval Academy in the celebrated James Sutton murder case. This case, which was either a suicide, a murder, or a duel, shook both the Academy and the Corps. The court eventually found no one guilty of murder, and recommended no further action should be taken against the individual officers involved in the incident; but Neville and the other members criticized the indiscipline of the young officers involved and the system of officer training.[33] Questions persist, however, and the truth of that murky, drunken night at Carvel Hall and on Worden Field may never be fully clear.

The U.S. intervention in Mexico in 1914 is now little recalled, except, perhaps, in Mexico. After years of the

Porfirio Diaz dictatorship, a faction led by Francisco Madero rebelled in 1910. Madero was elected president. But in 1913, Gen. Victorio Huerta arrested Madero, forced him to resign, and apparently had him assassinated. War erupted again, this time between Huerta's forces and Madero's supporters (including Pancho Villa). When Woodrow Wilson took office he refused to recognize Huerta, instituted an arms embargo,[34] and ordered a squadron of the Atlantic Fleet under Radm Frank F. Fletcher (USNA 1875) in Rhode Island to protect the Americans running the Mexican oil industry along the Gulf Coast.

But Huerta seemed to be stabilizing Mexico, and newspapers poked fun at Wilson's failure to teach the Mexicans, as Wilson put it, "to elect good men."[35] The "Tampico Incident", when Huerta troops arrested a paymaster and sailors ashore from Dolphin to buy gasoline, heightened tensions.[36] When the U.S. consul reported that a German ship, *Ypiranga*, was scheduled to deliver 200 machine guns and 15 million rounds of ammo to Huerta forces at Veracruz, the fuze was lit. Since seizing Germany's ship would mean war with her, Secretary of the Navy Josephus Daniels ordered U.S. forces to concentrate at Veracruz. In a 2:30 AM phone call Wilson issued the laconic order, "Take Veracruz at once."[37]

Lt. Col. Wendell Neville's Marines would be first ashore,[38] comprised of a battalion aboard *Prairie* and the fleet Marines from two battleships, totaling 22 officers and 578 men. His planning was pushed forward by an impending storm. At 11:12 AM on April 21, the first boatloads of Marines pushed off for Pier Four. They spilled out in khakis and campaign hats, knapsack rolls and Springfield '03's. Neville was assigned to occupy the Terminal, railway yard, cable office, and power plant. He led his men inland, setting up a main line of resistance west of the terminal station. By 11:45 all his objectives were in U.S. hands.[39]

Marines of Vera Cruz. Left to right: Captain F.H. Delano, Sergeant-Major John H. Quick, Lieutenant Neville, Colonel J.A. Lejeune, and Major S.D. Butler

So far so good, but shortly thereafter released criminals who had been armed by the retreating Mexican commander,[40] plus civilian militia and some regular Federalistas began first sniping, then machine-gunning the marines and seamen. Neville pulled his men back into warehouses, then set up Colt machine guns to dominate the streets. Then he went over to the attack, sending skirmishers to clear out the regulars.

Around one o'clock, *Ypiranga* hove in sight and was quickly persuaded to anchor rather than making up pierside. Meanwhile the cadets at the Mexican Naval Academy were firing a small cannon at boats landing reinforcements. Several 3-inch shells from *Prairie* killed one Mexican cadet and wounded another, and the resistance ceased. The invasion became a small-unit fight in the streets and across rooftops, with sailors and marines clearing buildings and snipers targeting the Americans.

During the night, most of the regular Mexican forces withdrew. The next morning, advancing bluejackets were fired on by civilians, including women. Neville extended his flank and began pushing south through house-to-house fighting. Around 11 AM Col. John LeJeune came ashore to take over command, but by then resistance was dying out. What Wilson had apparently intended as a peaceful occupation had killed over 126 Mexicans and 17 Americans. For his part in this two-day action, Neville received one of the amazing total of fifty-six Medals of Honor awarded for Veracruz.[41] Neville's citation read:

> In command of the 2d Regiment Marines, Lt. Col. Neville was in both days' fighting and almost continually under fire from soon after landing, about noon on the 21st, until we were in possession of the city, about noon of the 22d. His duties required him to be at points of great danger in directing his officers and men, and he exhibited conspicuous courage, coolness, and skill in his conduct of the fighting... His responsibilities were great and he met them in a manner worthy of commendation.

Following Veracruz, Neville sailed for China again, to serve at the Legation for two years, where he commanded the combined Allied guard.[42] He sailed from there for France in 1917, now a colonel, to take command of the Fifth Marines, and for a rendezvous with destiny that was to make all his actions to date seem like trivial skirmishes.

Elements of the Fifth Regiment had arrived in France in June, only two months after the U.S. declaration of war. The Secretary of War directed they be organized as infantry and attached to the Army expeditionary force, rather than operating independently.[43] Neville took command on January 1st.[44] At first assigned to provost duty with the AEF, the Fifth was pulled together and sent fifty miles back of the line in the Verdun region. They trained hard there for two months under the tutelage of the 77th French Infantry, which had been fighting the Boches since 1914.[45] The Fifth became the first Marine unit to occupy a front-line position in World War One[46] when they moved up as part of the AEF's Second Division. The sector had been quiet for some months, although fighting had been intense there in 1915.

But the Germans were on the march in the spring of 1918. All three Allied offensives in 1917—the French Nivelle offensive, the British at Flanders, the Italians at Caporetto—had been flung back with immense losses.[47] Now the Bolshevik collapse allowed the Germans to shift forces to the West. Ludendorff planned a massive attack on the strongest part of the enemy front, a "Great Battle" that would finally break the combat-weary British and mutiny-plagued French.[48] Prince Max of Baden asked the Oberst Quartermeister what would happen if the offensive failed. "Then Germany must perish!" was Ludendorff's reply.[49]

On March 21, 1918, the onslaught began with the most powerful artillery concentration in history.[50] The British line ruptured; the Germans surged forward. On March 23 the first shells fell in Paris. The Fifth was shelled, gassed, and raided, taking heavy casualties;[51] they had to learn fast in this searing crucible. Then Ludendorff shifted his axis of attack, first to the south, then to the north, overrunning Portuguese units. These thrusts were only barely held, but in one month the Germans had seized more territory than had either side in three years of agonizing war. "More than anything else, fear characterized the Allied position."[52]

After a short breather, Ludendorff developed another attack. This one would crash through to Paris and end the war. Plan Blucher would hurl three armies against the section of the line that held, coincidentally, the thirty-five miles of front along the Chemin-des-Dames that Pershing had stubbornly marshaled his Americans in. The attack

U.S. Marines in Belleau Wood, by Georges Scott

opened on May 27 and the Germans crossed the Marne in the first week of June,[53] reaching Château-Thierry, forty miles from Paris. Army divisions held them there and the Germans turned right toward Vaux and occupied the thick forest of Belleau Wood.[54]

Meanwhile the 2nd Division, which included the Marines, was rushed by forced marches and trucks up along the Paris-Metz highway. When they arrived, "they were gray with dust and hollow-eyed with fatigue. They looked more like miners emerging from an all-night shift than troops ready to plunge into battle."[55] They had no tanks, no gas shells, no flame-throwers;[56] their M1918 Chauchats were the worst light machine guns ever produced and tended to disassemble when fired[57]; but they were ready for what might be the most pivotal battle the Corps has ever fought. It was during this move up to the line that Neville was said to have been told by a retreating French colonel that he should retreat too. "Retreat, Hell," Buck supposedly said. "We just got here."[58]

Wendell Cushing Neville— courtesy of Nimitz Libary

Scores of books have been written about Belleau Wood. Pershing called it "... the biggest (U.S.) battle since Appomattox and the most considerable engagement American troops had ever had with a foreign enemy."[59] Suffice it to say that the square mile of heavy forest and large boulders was strongly held by the 461st Imperial German Infantry, heavy on machine guns and well supported by artillery.[60] The Marines assaulted across wheat fields into the forest on 6 June, an attack commemorated in one of the finest novels of combat ever written, Thomas Boyd's *Through the Wheat*.[61] This is the attack during which Gunnery Sgt Dan Daly asked his men, "Come on ya sons-of-bitches, do ya want to live forever?"[62] The back-and-forth struggle for this strategic position raged until June 26, with extremely heavy casualties, repeated attacks and counterattacks, gas, shelling, and hand to hand fighting. Not until June 26 could the message could be sent, "Belleau Woods now U.S. Marine Corps entirely."[63]

During these titanic events, the commander of the Fifth Regiment appears from the dust and smoke of battle only in glimpses. Neville comes across as an able tactician concerned above all with gaps in the line, but also worried for the care of his wounded.[64] Floyd Gibbons, correspondent for the *Chicago Daily Tribune*, caught him poring over maps on the kitchen table of a farmhouse on June 6, the day the regiment suffered its heaviest losses.[65] Neville gave him permission to go up to the line, but warned, "Go as far as you like, but I want to tell you it's damn hot up there."[66] On June 14, when things were not going well, Brig. Gen. James Harbord, U.S. Army, visited his headquarters after a bad gas attack. Out of the blue, Neville handed him a pair of Marine Corps collar devices and said, "Here, we think it is about time you put these on." Harbord later wrote, "I was as much thrilled by his brusque remark and his subsequent pinning them on my collar the next few minutes as I have ever been ... I wore those Marine Corps devices until I became a Major General, and I still cherish them as among my most valued possessions."[67] Whether dealing with generals or privates, all agree "Whispering Buck" Neville had the "common touch"—he wasn't a martinet, he inspired and cared for his men. Once, while his overcoat was hanging outside his tent to dry, an Army trooper unfamiliar with USMC insignia thought it was German, and cut off Neville's sleeve as a souvenir. Neville treated this as a joke at his expense. But five years after the war, when told that guilty doughboy was ill in hospital, he immediately ripped his insignia from his uniform and sent it to the sick soldier by registered mail.[68]

Following the battle, Neville was promoted to brigadier general. He was awarded an Army decoration, the Croix de Guerre with Palm, two regimental citations and the Legion of Honor. French Premier Clemenceau declared that July 4 a national holiday, invited the Marines to a gigantic parade in Paris, and offered to open the capital's brothels to the Americans. (General Pershing turned him down.[69]) Neville led the 4th Brigade in the Saint-Mihiel, Mont Blanc, and Meuse-Argonne offensives that ended the war.[70] Following the Armistice, he participated in the occupation of Germany and was also awarded the Army and Navy Distinguished Service Crosses, and two additional Croix de Guerres.[71]

Neville's ascent after the war was swift. He picked up his second star in 1920 and served as assistant to the

Major General Commandant Wendell C. Neville, Fourteenth Commandant of the Marine Corps, 5 March 1929–8 July 1930

Commandant,[72] including service on a court of inquiry into alleged crimes and offenses by the occupation force in Haiti.[73] Since Mrs. Neville had suffered badly during the flu epidemic of 1918, he next requested duty in a warm climate, and commanded the Department of the Pacific at San Francisco until 1927, when she died.[74]

When John LeJeune unexpectedly resigned as Commandant, he recommended Neville as his successor. Having fought side by side with now President Hoover in Peking cannot have hurt. Neville continued LeJuene's commitment to developing expertise in amphibious assault and reforming officer promotion. He had problems with Smedley Butler and with Haiti, as well as force reductions, pay cuts, and budget restrictions due to the Depression. The strain and his developing hypertension led to a stroke in March of 1930, and he lived only four more months.[75] Many old enlisted Marines walked in his funeral cortege at Arlington Cemetery, paying tribute to a man who was never the head of his class in smarts, but always at the right place at the right time with the right stuff.

8

MERIAN CALDWELL COOPER, CLASS OF 1915 (NOT GRADUATED): THE MAN WHO CAPTURED KONG

"The greatest American hero of the Twentieth Century"—that's what a foundation named after him calls Merian Caldwell Cooper.[1] Yet the name's unfamiliar to most of us. Film buffs may recall it, associated with the flickering image of a gigantic ape. But who remembers the pilot who brought Kong down from atop the Empire State Building?

Cooper attended the Academy, entering with the Class of 1915, but left under a cloud his first-class year. In some ways, his life from then on can be seen as an effort to redeem that failure.

But what impelled this "timid" boy to slip the surly bonds of the Naval Academy, for the sky? To become a true hero, if an unconventional one; to be decorated for valor in three wars; shot down twice, a POW of both the Germans and the Russians; and then, to celebrate heroism, exploration, and America in films that will live as long as images haunt our screens?

Merian Caldwell Cooper was born in Jacksonville, Florida, in 1893, the youngest of three.[2] His father was John C. Cooper, a well-connected[3] Democratic attorney and politician. Caldwell was his mother Mary's maiden name. Coopers had lived for generations near the Florida/Georgia border, and Cooper would retain for all his life the Lower South accent he grew up with there.[4]

"He was only 7 ½ years old when the Great Fire of 1901 roared through town and destroyed just about everything, including his family's home. This cataclysmic event left a big impression on the young boy. After the Fire, his family rebuilt their home one block away . . . . This was just about the time the silent movie industry was in its heyday in Jacksonville, and this exposure to the glamour and excitement of movie making also had a profound influence on him."[5]

"I was a l'il, timid boy," Cooper said of himself much later.[6] But the young Merian was nourished on tales of heroism. His great-grandfather, John Cooper, he was told, had been senior aide to the great Polish cavalryman and advisor to George Washington, Kasimir Pulaski, during the Revolutionary War. Family legend had Pulaski, mortally wounded near Savannah,[7] dying in Cooper's arms.[8] Though records don't seem to support the story,[9] it surely had an effect. Another influence was his great-uncle and namesake Merian Rokenbaugh Cooper, who'd enlisted in the Confederate army at sixteen, rose to captain, and was wounded in the Seven Days Battle.[10] Merian read tales of travel and adventure, including one of equatorial Africa with frightening illustrations of gorillas.[11] He built his stamina by swimming and working out, to prepare himself for the adventures he yearned to experience.[12]

Cooper secured an appointment to Annapolis through a family friend, a congressman. He reported to USNA in June, 1911.[13] But his tenure at the Academy was a disaster. He later ascribed his leaving to his telling the Superintendent that airplanes could sink battleships.[14] This seems to be not quite the way it happened.

If not all our sins are remembered, the Archives Division of Nimitz Library ensures our conduct infractions will be.

"Coops" in Jacksonville as a teen

"Alligator Joe"[15] was unsatisfactory in conduct for three successive years. The letters filed from his first class year are particularly blunt. His company officer: "I do not believe him capable of becoming of value to the Naval Service and of upholding its traditions and duties. Personally, I would not trust him with any important duty whatever." His battalion officer: "In my opinion Midn. Cooper is temperamentally unfit for retention in the Naval Service." The Superintendent: "There is no excuse whatever for such a discreditable record on the part of a First Classman." In December of his first class year, he began a period of confinement aboard the Yard's punishment ship.[16]

Cooper published a collection of impressionistic vignettes in 1927. Though marred by the casual racism of the period (perhaps why he later tried to buy up and destroy all the copies that had been sold), it's a look into what drove him. "Christmas 1914" recalls his being put in hack at the Academy.

"Someone switched on the electric lights. Twenty-odd young men lashing hammocks with the rapid precision of long practice. When these were stowed away the cabin showed Spartan bareness . . . It was the midshipmen's prison on the old wooden Spanish ship, Reina Mercedes, at Annapolis.

"The midshipmen there fell into double rank swiftly. Though my sleeve was bare of any distinguishing mark of rank, and though I had been stripped of the few small class honors I held and deprived of every privilege, nevertheless I was a first classman (the Naval Academy term for a senior) and so I took command. There were no other first classmen with me at Annapolis this Christmas. I was the black sheep of my class...."[17]

Cooper promised to reform, but didn't. Letters between the Superintendent and the senator who appointed the substandard mid make painful reading. The decision to expel him was made in February of 1915,[18] and he did not graduate with his classmates.

After his summary ejection from Annapolis, Cooper went through a crisis of confidence. He struggled with alcohol. "I was twenty-one. I had no training but that for a naval officer. I was in a strange city and almost penniless . . . Long and hard I hunted for a job—any kind of work—for I wished to eat. But for many days I found none. Money all gone. I pawned my suit-case for a dollar and a half. One dollar went for the weekly attic rent. I decided to economize with the remaining fifty cents . .. For five cents, in those happy days, one might get six large buns. I decided to live on six such buns a day. My plan was to eat one bun, then drink much water, in the hope the water would cause the bun to swell, and so partly fill my empty insides. Then another bun and more water, and so on again."[19]

Cooper tried to join the French or British air service, as the Great War was on, but failed. "After he was forced to resign from the Naval Academy he attended before the war, reportedly for his own anarchic behavior, Cooper set out to redeem himself in the eyes of his brother and father by enlisting to fight in Europe."[20] He worked as a journalist for several newspapers, but adventure still called. In 1915 he joined the Georgia National Guard. His unit was sent to Mexico after Pancho Villa, but he never saw combat.[21]

As the country came closer to war in 1917, though, Cooper got himself into pilot training. That October he went to France.

The 20th Aero Squadron was organized on 26 June 1917.[22] Trained in San Antonio and in Dayton on Jennies and Standards, the squadron sailed for the war zone that December. Meanwhile, Cooper was training at Issoudon, where he made the acquaintance of a Cedric E. Fauntleroy. After training, Fauntleroy went to the 94th Squadron, and Cooper to the 20th.[23]

In August 1918 the 20th was flying DH-4 Liberties.[24] The only U.S.-built plane to see combat in WWI, the underpowered, slow Liberties were nicknamed "Flaming Coffins."[25] Interestingly, Cooper was assigned a collateral duty along with his flying. Edward Rickenbacker called him the "official Movie Picture expert," and described how, using a captured German plane, he and Cooper recreated aerial combat "scenarios". In the first attempt, Cooper's plane crashed and was destroyed on takeoff, but he stepped out of the rear seat with both himself and his camera unhurt. Their next attempt looked so realistic that when they drifted over a nearby aerodrome while filming, they were targeted by a French antiaircraft battery. The exciting footage was spliced into newsreels and shown in France and the United States[26], giving Cooper his first film credits.

Cooper's first action came that September, when the section he commanded was ordered to fly recon flights over the German lines in preparation for the upcoming St. Mihiel offensive.[27] His squadron was hastily converted to bombers by adding racks under the wings, and flew in terrible rain and storms to support the attack. Losses were heavy, particularly as the squadron flew to support the Meuse-Argonne assaults.[28]

"On September 26, 1918, pilot Cooper and his gunner, Edmund C. Leonard, of the 20th Aero Squadron, took off in their De Havilland DH–4 biplane bomber from a muddy airfield in Amanty, France. Their mission was to cripple German troop and supply lines just prior to the critical Meuse–Argonne offensive. About an hour after takeoff and now over enemy territory near Verdun and Metz, Cooper and his observer-gunner were shot down by a German Fokker pursuit plane."[29] The official records of the Squadron state, "Lieutenant Merican (sic) C. Cooper who was brought down in this raid performed a heroic deed which won for him the Meritorious Service Cross."[30] Shot down, on his way to the ground in a burning plane, and without a parachute, Cooper was climbing out of the cockpit in order to shorten the dying when he noted his gunner was still alive. Despite gruesomely burned hands, he crawled back into the cockpit of the still-burning plane and managed to crash-land. Both airmen were taken prisoner.[31]

Cooper was carried as MIA until the Red Cross was able to ascertain his status. He spent the last months of the war in German hospitals and POW camps,[32] meanwhile being promoted to captain, and was released at the end of the war in November, 1918.

One would think that getting shot down on the Western Front, being badly burned and nearly killed, would be enough action for one lifetime. But Cooper had only begun his combat flying. Another war flared up as Germany withdrew from the newly-independent, or potentially independent, states created out of the fallen Russian and Austro-Hungarian Empires: Poland, Finland, Lithuania, Latvia, Hungary, and Ukraine, among others. Unfortunately, those bordering Russia had a new problem: the nascent Soviet Union. Poland, in particular, was in Lenin's sights. He would use ". . . .the bayonet to probe Poland's readiness for social revolution."[33] If Poland went Red, Germany could be next.[34]

President Woodrow Wilson, intent on not getting embroiled in Eastern Europe, even though he'd lit the fuze of nationalism there, refused U.S. help. Nor were Britain and France eager for more war. The new democracies were left to defend themselves, if they could.[35]

Following a reunion with some of his fellow pilots in Paris that Christmas, and his refusal to accept the Distinguished Service Cross he'd been awarded while the Army thought him dead,[36] Cooper was seconded by the Air Service to assist Herbert Hoover's American Relief Administration. The ARA sent him to Warsaw to coordinate delivery of food to Lvov, which was surrounded by hostile Ukrainian forces in a border conflict. Returning to Warsaw, Cooper requested duty in a theater in which he could fight Bolsheviks, for whom he seems to have generated a powerful loathing as a result of talks with captured Russians while a POW in Germany. He wrote that "America would one day have to fight them . . . we will be the only people in the world really worth plundering."[37] If the U.S. didn't want to take the Communists on, he was determined to fight them in the service of some other country, or even with the White Russian armies.

Sometime during this period, in Vienna, he met an ex-Army cameraman named Ernest Schoedsack. Schoedsack describes their first meeting: "I was at the Franz Josef Railroad Station. Down a platform came this Yank in a dirty uniform, wearing one French boot and one German one. It was Coop. He was just out of German prison and he wanted to get to Warsaw. He had once been kicked out of the Naval Academy and had sold his sword. Now he'd found the guy who had it and he'd bought it back."[38] Schoedsack was also headed for Poland, with the Red Cross, helping refugees escape and filming on the way.[39]

After obtaining the promise of a commission from Marshal Pulsudski himself, Cooper was detached from the Air Service in the summer of 1919, with the still-vague idea of forming an American squadron to fight for Poland.[40] Meanwhile, plans were also ongoing at the highest levels for a projected large American air-ground unit of Polish-Americans, modeled on the French Foreign Legion.[41]

At this point, at a Paris café, Cooper happened to run into Cedric Fauntleroy again. During the war, Fauntleroy had become a logistics specialist as well as a pilot. He, too, had been hired by the Poles, as an aviation technical advisor. Cooper pitched his plan for a squadron of ex-pat pilots, and Fauntleroy was instantly on board.[42]

There were other, higher-level players, of course: Pershing; the State Department; the French; President Paderewski, and the Polish High Command. Eventually the idea of a ground force was abandoned; but that of an air squadron survived. Perhaps there was an element of deniability—if only volunteer pilots were involved, paid and supplied by Poland, not the U.S.

Regardless, Cooper and Fauntleroy set to work. They were looking for idealists, not avaricious mercenaries. Airmen would cover their own transportation costs, and accept only the minuscule salaries of regular Polish officers. The first nucleus of pilots left Paris in unmarked uniforms, in an American Typhus Relief train headed east.[43]

The new air force was makeshift at best, made up of pilots from Austria, Russia, Turkey, and other countries, and equipped with whatever could be found, scrounged, or rebuilt from the wrack of a world war. The Americans were assigned to the 7th Fighter Squadron at Lvov. Arriving in October 1919, Fauntleroy took command and requested redesignation into the "Koscuizko" and "Pulaski" squadrons.[44] The former name took hold, and the pilots began training on planes supplied from hangars full of abandoned equipment.[45] The Eskadra Kościuszkowska was born, mainly flying the war-surplus Albatros D.III.

The D.III model Albatros "was introduced early in 1917 and it met with instant acceptance by the German pilots. It was easy to fly and was an effective combat aircraft . . . . Initially, the narrow lower wing was susceptible to frequent failure in prolonged dives, but with reinforcement of the structure and improved workmanship, the problem was ameliorated. The Albatros D.III served with great success throughout the first half of 1917."[46] By 1919 they were obsolete, but still good enough for ground attack. (Scale-model enthusiasts: Encore has a model of Cooper's "Koskiuzsko Squadron" Albatros that you can build.)[47]

The specifics of the complex and vicious Russo-Polish war are too involved to recount here.[48] But as the battle lines burned back and forth across Eastern Europe through 1919 and 1920, the Americans strafed and bombed Bolshevik cavalry and troops, shot up locomotives, bombed bridges, interdicted river traffic, and provided reconnaissance. Fauntleroy and Cooper repositioned the squadron via road and rail transport to help meet and blunt enemy offensives, a foreshadowing of later developments in close support of ground operations. Their D.IIIs, and, later, Italian Balilla biplanes, marked with the Polish chessboard and a distinctive squadron insignia that combined US and Polish symbols, became a familiar sight above the battlefield. So familiar, that Bolshevik writer Isaak Babel included a reference in his short story collection, *Konarmiya*: "And Trunov pointed to four dots in the sky, four bombers that came floating out from behind the shining, swanlike clouds. These were machines from the air squadron of Major Fauntleroy, large, armored machines . . . the major and three of his bombers proved their ability in this battle. They descended to three hundred meters, and first shot Andryushka and then Trunov. None of the rounds our men fired did the Americans any harm."[49]

Babel's work makes clear that this wasn't just a war for territory, but for ideology. The foreign capitalist pilots could expect no mercy from those they strafed. Cooper carried a vial of poison on each flight, in case he was taken prisoner.[50]

Cooper (1920) from the Hoover Institute

On July 13, the inevitable happened. Attacking a Cossack force from low altitude, Cooper was shot down again. He had no chance to commit suicide: he blacked out after the crash, and was captured.[51]

The Cossacks whipped him and staged a mock lynching, but Cooper was saved by presenting himself as a humble enlisted man, "Frank Mosher", whose name was stenciled on the secondhand long underwear he was wearing.[52] The ruse worked, even in an interview with Semyon Budyonny, whose train the squadron had passed up attacking several times, knowing his wife traveled with him.[53] Cooper also may have been interrogated by Timoshenko, or even Stalin himself,[54] who was with that army, and was offered a position with the Bolshevik Air Squadron. He refused, tried to escape, was recaptured, and sent to Moscow under heavy guard.[55] Meanwhile, that August, "Flying a total of 190 sorties, dropping nine tons of bombs, Polish and American airmen managed to slow Budyonny's advance to only a few miles a day, buying precious time for Polish land forces to move to counter the Soviet threat."[56] To the world's surprise, Polish forces held Warsaw, repulsing the Soviets so decisively they retreated.

Imprisoned again, Cooper suffered through the "Starving Winter" of 1920–21, when peasants cut off food to Moscow. However, he managed to contact Marguerite Harrison, a newspaperwoman and socialite then in Russia

Model of Cooper's Polish Fighter Plane

both as a reporter and as a spy for U.S. Military Intelligence.[57] He reminded her that they'd danced at the Hotel Bristol in Warsaw, and asked for help. Harrison got word back to his family that he was alive, and arranged food parcels that probably saved Cooper's life.[58]

After being moved to another prison, Cooper escaped along with two Polish officers. He pretended to be mute during an arduous, dangerous, 800-kilometer march to Latvia.[59] At one point he killed a Red Army soldier, cutting his throat with a knife.[60] Cooper later credited his survival and escape to USNA, writing, "I take no credit, but credit only the tough training I had at the United States Naval Academy at Annapolis.[61]

An armistice was signed between Poland and the USSR that October, followed by negotiations to settle the border question (temporarily, as it turned out.)[62] After his release, Cooper also found time to father an out-of-wedlock son by Marjorie Crosby-Slomczynska, an English-Russian emigrée. Maciej Slomczyński grew up to fight in the Polish Underground in a later world war.[63]

Cooper was discharged from the Polish military in 1921, with the rank of lieutenant colonel. Along with several other pilots, he was awarded the Virtuti Militari, Poland's equivalent of the Congressional or Victoria Cross.[64] Interestingly, he does not seem to have turned this one down, as he did the Distinguished Service Cross.

As the dust settled in Europe, Cooper continued to seek adventure. He worked the crime beat for the *New York Times*, then joined a film expedition to the South Seas and Africa with Edward Salisbury, a millionaire explorer and filmmaker.[65] When the cameraman quit, Cooper remembered Schoedsack, and the ex-Signal Corps cameraman joined the expedition. *Wisdom* sailed through the Andamans, Sumatra, the Marquesas, the New Hebrides, the Solomon Islands, and across the Indian Ocean to the Red Sea. Cooper met headhunters, lepers, cannibals, "beast men," and Emperor Haile Salassie. Grounded off Yemen, they prepared to hold off pirates with Cooper's Thompson gun.[66]

After writing *The Sea Gypsy* with Salisbury, Cooper decided to make his own film, combining what today would be called a travel documentary with the adventure of exploring new lands and peoples. (He called later efforts "folklore picture tales."[67] or "natural drama."[68]) Fascinated by Iran, he and Schoedsack decided to document the grueling annual migration of a nomad tribe in search of forage for their animals. Contacted for funding, Cooper's rescuer in Moscow, Marguerite Harrison, kicked in some cash, but also insisted on coming along.[69]

"For six months, the "production company" lived with the Bakhtiari and photographed every stage of their trek to reach grass . . . (afterward) Cooper and Schoedsack went to Paris, and there, because of lack of money, developed and printed the entire 50,000 feet of exposed negative themselves. When they arrived in New York (Mrs. Harrison went there directly from Iran) Cooper and

Schoedsack rented some equipment and a small room and edited the material that was to become the documentary entitled *Grass* (1925)."[70]

Though not a huge moneymaker, *Grass* became one of the most famous films of the silent era.[71] Cooper and Schoedsack followed it with a better-financed film about trapping man-eating tigers in Siam. One of the high points of this expedition was when they tried to transport a captured man-eater named Mister Crooked in a makeshift cage in a native canoe. That night, during a raging storm on the river, the tiger began dismantling the cage from inside. Our heroes solved this dilemma by pouring chloroform into the tiger's mouth through a piece of bamboo. Cooper wrote, "It did not put the tiger out, but after we had repeated the dose a couple of times it sort of discouraged him. Indeed, it discouraged him long enough to give us time to repair the cage. And when he recovered and tore a log or two loose again, we repeated the performance."[72] Cooper's expert rifle marksmanship saved several people during the filming.[73] Released by Paramount in 1931, *Chang* was a box office hit.[74]

Their next project was *The Four Feathers*, about a British officer who seeks redemption after a cowardly act. The outdoor scenes were filmed on location in East Africa. The Hollywood part of the production starred Richard Arlen, William Powell, and a riveting young ingenue, Fay Wray. Their producer was a young David O. Selznick.[75]

Meanwhile, Cooper became bicoastal, involved in commercial aviation financing in New York, along with Juan Trippe, John Hambleton[76], and C.V. Whitney, who established Pan American Airways. 1n 1927, Cooper had put most of his movie profits into aviation stocks. By the early thirties, he was a director of Pan American, Western Airlines, General Aviation (forerunner of North American), and other airlines.[77] Along with his executive duties, though, Cooper had begun to sketch out a fictional project about a filmmaking explorer who brings a gigantic gorilla to New York.

New York Daily Mirror *Cartoon*

The idea probably stemmed from many sources: a glimpse of a plane cruising over the Empire State Building, Cooper's documentary filmmaking, his voyages to mysterious islands, his flying . . . and the idea, presented in different ways in his previous films, of the conflict of civilization with a rapidly-vanishing Nature. The project he first called *Kong* was born.[78] Edgar Wallace contributed

Schoedsack (L), Jean Arthur, and Cooper (R) in Four Feathers

the first draft of the screenplay, which contains some of the elements we recognize, though he died before shooting began. Ruth Rose, a naturalist and Schoedsack's wife, completed the script.[79,80] Another major influence was Willis O'Brien's previous stop-motion work on a film to be called *Creation*.[81] Cooper actually came back to RKO to straighten out *Creation*, which was seriously over budget and had other problems, but realized he could transform its scenes and models, with O'Brien's cooperation, into his gorilla movie.[82]

Entire books have been written about *King Kong* (1933), and it's been interpreted in dozens of ways. It blends ancient archetypes and folk tales with major advances in stop-motion animation, innovative special effects, a taut script, and riveting acting. We will note only that you can actually see Cooper at the end of the film; he and Schoedsack man the biplane that shoots Kong off the Empire State. Cooper reportedly said, "We made him; we should kill the sonofabitch ourselves."[83,84] The planes were Navy Air, from Floyd Bennett Field in Brooklyn.[85]

The same year, Cooper married actress Dorothy Jordan. During that decade, he wrote, directed, or produced numerous films for RKO, including *The Last Days of Pompeii, Roar of the Dragon, Headline Shooter, Lucky Devils, The Phantom of Crestwood, Flying Devils, Son of Kong, Rafter Romance, Double Harness, The Right to Romance, One Man's Journey, Living on Love, The Most Dangerous Game, A Man to Remember, She*, and doubtless others.[86] When Selznick left the studio, Cooper took over as production head of both the RKO lot in Hollywood and the Pathé studio in Culver City. Movies made under his overall supervision include *Morning Glory, Melody Cruise, Professional Sweetheart, Ann Vickers, Ace of Aces, Little Women, Flying Down to Rio*, and *The Lost Patrol* (directed by John Ford).[87] But eventually, frustrated by RKO's unwillingness to make pictures in color, a heart attack, and creative differences, Cooper left for MGM and for Pioneer Pictures, which he founded with Jock Whitney and David Selznick.[88,89] "He was instrumental in Selznick's Technicolor production of *The Garden of Allah* in 1936 that paved the way for Technicolor's final acceptance by the public with the filming of Selznick's *Gone With the Wind* in that process in 1939."[90]

As World War Two started, Cooper had been working on what today would be called heroic fantasy, about a publicly humiliated Air Corps pilot who crashes on an undiscovered island, where giant eagles are tamed for use in battle. *War Eagles* was to have surpassed *Kong* in special effects, but it was overtaken by events and never made.[91] Instead Cooper, who had retained his reserve commission, went back on active duty in June, 1941.[92]

Before the U.S. entered the war, Cooper reported to Hap Arnold on the German bombing of London. After Pearl

Harbor, he helped plan "Project Aquila," a raid on Japan. Aquila was originally two-pronged, with B-17s from China attacking simultaneously with Army B-25s launched from Navy carriers. Cooper essentially served as exec for the China-based force of B-17s and B-24s, but even as Doolittle's carrier-based raid succeeded, Army mismanagement and lack of coordination, plus a vicious Japanese preemption, derailed that attack.[93]

Instead, Cooper hooked up with the American Volunteer Group, the famous "Flying Tigers" (later the China Air Task Force, or CATF, based in China). Claire Chennault's take on Cooper is vivid: "My first chief of staff, Colonel Merian C. Cooper, was a character straight from the Hollywood movies he once directed . . . with his shirttails flapping in the breeze, a tousled fringe of hair wreathing his bald spot, a mantle of pipe ashes over his uniform and sagging pants, Cooper would never have passed muster at a West Point class reunion but he was a brilliant tactician and a prodigious worker . . . when planning a mission for the CATF. Cooper worked around the clock until every detail was satisfactory and then rode the nose of the lead bomber peering over the bombardier's shoulder at the target."[94]

Cooper planned and flew numerous successful bombing missions, masterfully employing disinformation to misdirect the Japanese. "Colonel Cooper had made the rounds of the Kweilin cafes the night before, dropping discreet indiscretions about the terrible shellacking scheduled for the Japs at Hong Kong. The C.A.T.F. formation headed straight for Hong Kong. But at the last minute the bombers cut sharply toward Canton and caught the Japanese flat-footed."[95]

A 1943 *Shipmate* article reported, "The raid over Canton was the most successful carried out by the China Air Task Force. Twenty-nine Jap pursuits were shot down; according to a Chinese dispatch 42 Jap planes were destroyed on the ground; and two ships were sunk. While some planes were shot and some men were wounded, the American suffered no losses in this raid. Colonel Cooper had a narrow escape during this raid. Colonel Robert Scott, pursuit commander, shot down a Jap which was too close to ""Coop's" plane."[96]

But differences escalated between Chennault and General "Vinegar Joe" Stilwell, based on disagreement about political and military strategies in China.[97,98] An additional factor was Cooper's (as it turned out, prescient) fear of an eventual Communist takeover there. This was exacerbated by a frank personal letter Cooper wrote to his friend "Wild Bill" Donovan, which Donovan circulated around Washington,[99] and another letter, over Stilwell's head, directly to President Franklin Roosevelt via Wendell Willkie that Cooper and Chennault co-wrote.[100] Mark Cotta Vaz says "Cooper took the fall" for Stilwell,[101] and he was returned to Washington on grounds of ill health. But back in DC, the still-outspoken colonel repeated his conviction that Stilwell was opening the way to a Communist victory. As he later recounted it, a long, contentious closeting with George Marshall closed off any possibility of further promotion.[102]

Shipmate described Cooper as he appeared then: "He gained 27 pounds from late December when he returned to Washington and his family until mid-February. But he still could play the role of the "The Thin Man" for one of his movie colleagues ... Today Colonel Cooper is energetic and a lover of action. When he talks he walks up and down, sits with feet on the desk, knocks the ashes from his pipe, refuels, lights up again, and walks some more. He reels off facts and figures about war in China and the far-east from his fingertips. He itches for his next assignment."[103]

Returned to duty in New Guinea, Cooper served as chief of staff to General Ennis C. Whitehead, General George Kenney's deputy commander. "Cooper helped guide the New Guinea invasion, the first completely airborne invasion made by American troops. Later still, Cooper was deputy chief of staff for all the Air Force units in the Pacific under General Douglas MacArthur."[104] He was no armchair planner there, either; between May and August 1943 he went along on at least nine combat missions.[105] But either chronic dysentery or Marshall stalled further promotion. His recommendation for the Distinguished Service Medal read, "Colonel Cooper was a key figure in the destruction of Japanese air power in New Guinea, which cleared the way for the occupation by Allied Forces of the whole north coast of that island." Cooper was among those few invited to witness the Japanese surrender aboard Missouri in Tokyo Harbor.

Following the war, Cooper returned to film, but to Westerns, not fantasy or travel adventure. Teamed with John Ford in Argosy Pictures, they made *The Fugitive* (1947), *Fort Apache* (1948), *She Wore a Yellow Ribbon* (1949), *Rio Grande* (1950), and *Wagonmaster* (1950).[107,108] Along with providing entertainment (and making an icon of John Wayne), Cooper wrote that he intended these films to counter Communist propaganda, to promote freedom and the American way of life.[109] He also produced *Mighty Joe Young*, *The Quiet Man*, and *The Sun Shines Bright*. In his later years he worked to advance Cinerama, a wide-screen viewing experience, and continued to be active in wealthy anticommunist political circles.

Cooper always maintained his links with the Air Force. He was promoted to brigadier general in the Reserve in 1950. In 1952, he received an honorary Oscar from the Academy of Motion Picture Arts and Sciences "for his many innovations and contributions to the art of motion pictures."[110] He died in 1973, after a long battle against cancer.[111] He was survived by Dorothy, who died in 1988, and their children, two daughters and a son.[112] The son, Richard Cooper, graduated from the Air Force Academy, commanded the 412th Test Wing, and retired as a colonel.[113]

Cooper's Star on the Hollywood Walk of Fame

In 1943, Cooper told *Shipmate*, "I was not a good midshipman. But the Naval Academy did a lot for me. Perhaps I wasn't good for the Naval Academy, but it was good for me."[114]

Merian Cooper demonstrated incredible courage, not least in returning to combat flying again and again despite burns, many crashes, and two spells in POW camps. He also displayed tremendous organizational ability, serving as chief of staff to an air force at war. Finally, if we credit his (perhaps dramatized) version of his interview with Marshall, he displayed moral courage to the point of career suicide.

What, then, did Cooper observe at USNA that persuaded him that he "wasn't good for the Naval Academy"—that a naval career wasn't worth restraining his wayward instincts? Conversely, what did the Administration see that persuaded them he could not be useful, or even tolerated, within the System? That his superiors could "not trust him with any important duty whatever"?

It does seem, as we pursue this series of biographies, that often those who were most badly behaved as mids won acclaim in later life. Certainly such individuals as Cooper, McGiffin, Evans, Nimitz, and many others were outliers according to the USNA disciplinary system. Is, or was, that system too intolerant of transgressive behavior that did not involve actual dishonesty or violence? Or is it simply unwise to try to confine such individuals . . . like massive, dangerous, and ultimately destructive monster gorillas . . . within the cages of military civilization?

Food for thought, indeed.

David Poyer

9
HOWARD W. GILMORE, CLASS OF 1926: "TAKE HER DOWN"

*The Devil smiled, "That's Death up there in the periscope shears,
He came here to help me turn your bravery into tears.
Death," said the Devil, "Go see how many you can drown."
It was then that Gilmore commanded, "Take her down."*
—Bill "Radar" Hagendorn

The first American submariner to win the Medal of Honor died with a phrase on his lips that's still a Navy byword. *Stag* magazine called him "tall and jut-jawed," but he was no Harrison Ford. His chin was receding. His hairline was heading north. He was an average guy, a quiet boy, a "down-to-earth man who got along with everybody."[1]

There's no denying Howard W. Gilmore was a genuine hero. But beyond that, something about the man remains elusive. A meek, Penrod-like face peers out of high school photographs. A pudgier, rumpled version confronts us from wartime photos. Competent, yes. Intelligent—very. But Gilmore had none of the devil-may-care panache of Philo McGiffin, the rigid pride of Richmond P. Hobson. No legends have grown up around him. Time has erased everyone who knew him ... including one very old former sailor, a former shipmate aboard the boat he died saving.

To him alone was Gilmore still a living memory, when he was interviewed for this chapter.

Howard Walter Gilmore was born in Selma, Alabama, in September of 1902.[2] But it seems he wasn't very attached to that town, or even that state. The census of 1900 records Gilmore's parents, Walter and Vernon, as living in Meridian, Mississippi. They may have been in Selma only a few years; no directories for that city survive from that period except for 1909, and the Gilmores aren't listed then. However long they stayed, by 1910 they were back in Meridian, with both Howard and a younger son, Francis.[3] The census gives Walter's occupation as "dry goods dealer," in another words, a clothier. Howard also attended school in Texas, but he actually seems to have considered himself from Louisiana; that's how he signed himself in his application to the Academy,[4] and the 1926 *Lucky Bag* lists him as being from New Orleans.[5] Clearly, this was a young man who hopscotched around growing up.

Gilmore's Meridian high school yearbook seems to picture a lad who was liked, but who stood out neither for fecklessness nor in any leadership role. He stares out from his photo with that receding chin, slicked-down, neatly parted hair, and high stiff collar, alert but not exactly a commanding presence. More than anyone else, he looks like Alfalfa, played by Carl Switzer, in the old "Our Gang" comedies. The quotation his classmates awarded him bears out this lack of impression: "Who never said a foolish thing, And never did a wise one." His clubs were the Hi-Y Bible Club, the French Circle; the Prentiss Literary Society. The Hi-Y lists him as a member, but not an officer. The Prentiss Society, again, lists him as a member, but not as an officer.[6] Gilmore may also have attended Ball High School in Galveston, but that school's records from that period are incomplete and shed no light on his time there.[7]

Gilmore's entry in his Meridian high school yearbook

Boys' Hi-Y Bible Club page from the 1919 Kaldron

Gilmore must have enlisted in the Navy the fall after graduating from high school. He was a yeoman second class in San Francisco when he took the competitive exam for the Naval Academy, and received an appointment-at-large from the Secretary of the Navy.[8] After three months at the Naval Academy Prep School reviewing quadratics, algebra, and other subjects, he was discharged from the Navy on June 20, 1922, and sworn in as a midshipman the next day at the age of 19.[9]

Gilmore's Academy career was neither spotless nor dreadful. He was neither a standout nor a bottom-scraper, though he did manage to earn academic stars and was obviously talented in math and engineering (rather than dago and bull). His monthly plebe year records, held in the USNA Archives at Nimitz Library, show wildly varying marks. By the end of plebe year, he ranked 223rd in aptitude and 23rd in "general merit" out of the 733 members of his class, making up by his grades what he lacked in accumulating nineteen demerits in the deportment department.[10]

It's difficult to read old midshipman records; some of the terms we can interpret, others we can't; their meanings have been lost. Through a glass, darkly, we seem to glimpse a man who achieved academically, but who otherwise stayed in the middle of the pack until his senior year, when several things happened at once. He was hospitalized for acute appendicitis in June of 1925, but recovered quickly enough to be retained at the Academy for aviation training for second class summer. Meanwhile, the red-haired, freckle-faced Gilmore picked up the nicknames "Gil" and "The Count."[11] Early in his Firstie year, though, he got into some unspecified but major trouble for which the Commandant stripped him of his Midshipman Officer rating. His 1/c conduct record also shows Class A paps for improper use of tobacco, conduct to the prejudice of good order and discipline, and improper performance of duty. Gilmore still graduated 34th of a class of 457, though, since his high grades outbalanced his heavy demerit load. He was carried on a list for flight training right up until March of his Firstie year, but apparently failed the physical; his health record is stamped "Not Physically Qualified" for any duty involving flying.

Gilmore in the 1926 Lucky Bag

With aviation training closed to him, perhaps because of bouts of chronic appendicitis, Gilmore chose battleships—the smart move in those days. His peacetime career is quickly recounted in the spare phrases of a Navy press release. He served for three years aboard *Mississippi* (BB-41), then nine years old and the pride of the Fleet, operating off the Pacific coast and sailing into Caribbean and Atlantic waters during the winter months.[12] After six months aboard *Perry* (DD-340) after her recommissioning, Gilmore apparently caught the submarine bug. He never got to write his memoirs, but if he had, might have echoed George Grider, who followed him from Mississippi to destroyers to submarines. "We worked with submarines . . . the first Navy experiments with sonar. The work was fascinating, and the life of the submarine men themselves fascinated me even more. I rode with them when I could, and soon I decided that was the life for me: small, happy ships with wonderful morale, wonderful spirit, and wonderful pay."[13] After graduation from Submarine School in New London, Gilmore joined

S-48 at Coco Solo in the Canal Zone, serving for a little under a year under an obscure but demanding executive officer named Hyman Rickover.[14]

He returned to Annapolis in 1932 for postgraduate instruction in ordnance engineering there and at the Washington Navy Yard for three years. In June of that same year he married Hilda St. Raymond in New Orleans.[15] After that he went to Groton to fit out and commission *Shark* (SS-174) at the Electric Boat Company. He served as executive officer and navigator from her commissioning until late 1937, when he joined *Dolphin*. Gilmore probably met Admiral Charles Lockwood aboard *Shark*, since Lockwood commanded SubDiv 13.[16] One of the more noteworthy incidents in his life also occurred when he was aboard *Shark*; on liberty ashore in Panama, he was set on by a group of thugs, who literally cut his throat.[17] (Fortunately, he survived the attack and recovered.) From there he went to shore duty at Dahlgren. With war clouds gathering, he took command of his first boat, *S-48*, the same craft he'd

S-48 in Coco Solo, 1931

served on with Rickover, when she was recommissioned in early 1941. He was providing services to submarine and antisubmarine warfare training commands at New London and Portland, Maine, homeporting out of New London, until five days before the war broke out.[18]

The submarine Gilmore fitted out at Electric Boat, *Growler* (SS-215) became one of the most famous fighting subs of World War Two. She was commissioned on March 20, 1942, a little over four months after the Japanese attack on Pearl Harbor.

Norman Friedman calls *Gato*, her class leader launched in 1940, "the prototype of the mass-produced 'fleet boat.'"[19] The design grew from a long and bitter struggle between proponents of the large "fleet" versus smaller "coastal defense" submarines, personalized by the conflict between Lockwood, now head of the Submarine Desk in Washington, and Admiral Thomas Hart, chairman of the General Board.[20] Fortunately neither side won, and neither completely lost. The compromise design combined long range and compactness, firepower and seakeeping ability, in order to accompany the Fleet across the Pacific to execute War Plan Orange.[21] *Growler* was fourth to be launched of a class that eventually numbered eighty-one boats.[22] The Gatos were larger than contemporary German designs (VIIAs) and had nearly double the range. But they were smaller than Japanese fleet boats (KD7s), and much more maneuverable. With a range of up to eleven thousand miles, surfaced speeds of up to twenty knots, ten torpedo tubes, lightweight diesel engines, and sophisticated (for the time) sonar and fire control computers, they were deadly weapons. With such features as air conditioning, fathometers, radar, pit logs, peloruses with night lighting, ice-cream freezers, movies, record players—luxuries unimaginable to "Sugar Boat" crews[23]—they could operate independently for long periods of time, deep in enemy waters.[24]

And in 1942, after shakedown training, that was where *Growler*, and Gilmore, were ordered to go.

"They told me to go out there and raise hell," said another skipper gaining command at about the same time.[25] With the destruction of the heaviest units of the Fleet on December 7, the mission of the sub force had changed from scouting to unrestricted warfare.[26] *Growler*'s first assignment on arrival in the Pacific was to guard Hawaii as one of seven subs assigned to picket duty during the Battle of Midway (4–7 June 1942). After this she sortied on her first war patrol.

Gilmore refueled at Midway and headed north, to the Japanese-occupied Aleutian Islands. On station off Kiska on 5 July, Gilmore began his combat career with an astounding action of nearly incredible aggressiveness, coolness, and professionalism: he attacked three anchored Japanese destroyers single-handed. After a cautious approach, he

USS Growler *patch*

Growler *in 1942*

Growler *in 1943*

Growler *in 1944*

fired two torpedoes at the first two, scoring hits amidships on both. He then fired two more fish at *Arare*, which blew up after firing two torpedoes at *Growler*, which dived to one hundred feet as the enemy torpedoes went by on either side, close enough to be heard by the crew without sonar. A Japanese hunter-killer group pursued him, but Gilmore skillfully evaded and later surfaced to find smoke billowing up where his prey was listing and sinking. Arare sank and the other two had to be towed

Gilmore in khakis

back to Japan for repair.²⁷ Gilmore brought *Growler* back to Pearl on July 17 only slightly damaged, and was awarded the Navy Cross for this aggressive action.²⁸

Growler sortied again a month later after repairs. She refueled at Midway again, but this time headed west, toward Formosa (Taiwan), then part of the Japanese Empire. Ranging in close to shore, Gilmore sank a heavy gunboat, two cargo ships, sampans, and a large supply ship. He expended his last four torpedoes attacking a large convoy, but missed with all four. Pursued by patrol boats, he ran through the convoy at periscope depth to lose them before terminating his patrol and returning to Pearl.²⁹ This second patrol, with a total score of 25,946 tons of enemy shipping, brought him a Gold Star in lieu of a second Navy Cross.³⁰

After a refit, Gilmore went to sea again in October for the sea lanes south of the main Japanese base at Truk. He sighted several contacts but was unable to gain an attack position. After a December refit in Australia, he sortied again on the fourth war patrol—the one which was to become legendary.

After several sightings off New Guinea that could not be developed for attack, Gilmore engaged a heavily escorted eight-ship convoy and sank a passenger-cargo maru. Evading a depth-charge attack, he resumed the patrol. He fired three torpedoes at a cargo steamer on 30 January, but

"Smack the Japs" poster ca. 1942

missed. He attacked a patrol boat on 31 January and two others on 5 February, in poor visibility, but without result other than a barrage of depth charges that blew out gaskets and started a serious leak in the forward torpedo room. Unable to surface because of the enemy patrol boats still searching for him, and taking a thousand gallons an hour from the steadily increasing leak, Gilmore finally broke contact by heading west, surfaced after dark, and got the leak under control.[31]

Most submariners in World War Two were wary of warships. Their preferred targets were cargo vessels, both for strategic reasons and because they were less dangerous targets. Howard Gilmore seemed never to hesitate about attacking enemy combatants, though, no matter how heavily armed or what the odds. On the night of February 7, that aggressiveness was to result in his death.

The night was overcast and very dark. *Growler* was operating on the surface northwest of Watom Island in the East New Britain Province, west of the Japanese base at Rabaul, when the lookouts sighted a ship on the starboard bow, on the opposite course, about a mile distant. Gilmore swung away to ready his tubes, then back to close. Night and poor visibility, though, gave the enemy gunboat a chance to reverse course without the conning officer detecting that the range was now closing fast.

Suddenly, at 0134, the range was too close to fire torpedoes. Like Ensign John F Kennedy six months later in *PT-109*, *Growler* was caught by surprise on the surface by an onrushing enemy with alert lookouts, manned guns, and a skipper as aggressive as her own. "Left full rudder" came down from the bridge. Then the collision alarm sounded.

At 0135 *Growler* T-boned the enemy ship while swinging to the left at seventeen knots, hitting him halfway between the bow and the bridge. The impact knocked everyone down and heeled the sub fifty degrees. The enemy opened fire with heavy machine guns at point-blank range, sweeping the conning tower. Gilmore ordered, "Clear the bridge." The quartermaster and officer of the deck got below, followed by two wounded survivors pulled down through the hatch.

The patrol report, submitted by the executive officer, Lcdr. A. F. Schade, USNA '33, continues: "About 30 seconds elapsed. No one else appeared at the hatch. Sounded diving alarm, closed the hatch. Submerged.

"The Commanding Officer, Assistant O.O.D. and one lookout were left on the bridge. It is believed they were killed by enemy machine gun fire and washed overboard."[32]

It seems worthwhile to more closely visit the question of when and how Gilmore actually gave the famous order attributed to him. At that time, a diving order typically came from the bridge in the form of a klaxon operated by push button, not a spoken word.[33] George Grider, commander of *Flasher*, reports a tradition during the early years of WWII of spicing submarine patrol reports with "color and a dash of bravado . . . there are the newspapers and the history books to be considered, and the long tradition of terse, crackling epigrams that adorn the nation's naval records."[34] But the Patrol Report makes no mention of such an order. On the other hand, the Medal of Honor citation states quite definitively that "Struck down by the fusillade of bullets and having done his utmost against the enemy, in his final living moments, Comdr. Gilmore gave his last order to the officer of the deck, "Take her down.""[35] A June 1943 article in *Shipmate* states that Gilmore issued this order, not via klaxon or IC or yelled down the hatch, but directly to the OOD, who, it will be remembered, then made it below, presumably passing the order along to Schade.

A 1963 article in *Stag* magazine, though, obviously based on an interview with then-Rear Admiral Schade and featuring a photo of him, not Gilmore, quotes the order as "Take her down, Schade! Take her down!" and specifically says the order was yelled down the hatchway after the wounded men and the QM were safely on the ladder—that is, after the OOD must have been below. Then the hatch banged shut. Schade, bruised and shaken from being knocked off the ladder back down into the control room by the collision, passed the order to the men at the diving stations. "Tears streaked down Schade's cheeks as he repeated the order."[36]

It probably happened that way. Or Gilmore may have told the OOD simply to press the klaxon button. Be that as it may, *Growler* tilted, and the tattoo of bullets ended as the conning tower slid under, water streaming in through bullet holes in the pressure plating. All gyro, internal comms, lighting, and heater circuits flooded out. The TBT (necessary for visual torpedo firing), bridge gyro, sonar, the SJ radar, both periscopes, and other equipment were shot away, smashed, or flooded out. There were six

Shipmate *art from June 1943*

inches of water in the control room and several feet in the pump room.

Despite this damage, and continuing leaks, half an hour later Schade took her back up in a battle surface, ready to fight for his commander. They found the dark surface empty; no enemy gunboat (presumed sunk in the collision, but actually the 900-ton *Hayasaki* survived[37]); no Gilmore; no Ensign William Wadsworth Williams, the assistant officer of the deck; no fireman third class Wilbert Fletcher Kelley, a lookout. Schade cleared the area to the west, but *Growler* was severely damaged by the collision and fusillade. Eighteen feet of her bow was bent back at right angles, the bow buoyancy tank was crumpled, the tube doors were jammed open, and a fire broke out in the maneuvering room. But the crew turned to and *Growler* was able to submerge at dawn, still leaking, but under control. Schade assumed command and surfaced at

Illustration from Theodore Roscoe's book

dusk to report the damage. Seven minutes after midnight the next day, Commander Task Force 42 directed her return.[38] Gilmore and Ensign Williams were reported as missing in action, as it was at least possible they'd survived.[39] Kelley was reported killed and "consigned to the sea." But none was ever seen again.

Despite her crumpled bow, *Growler* made it to Brisbane, where Schade and the crew were commended for bringing her back.[40] (Schade received the Navy Cross.) After extensive repairs, she made four more war patrols under his command. Three patrols later, she was lost attacking a convoy off the Philippines in October of 1944.

Gilmore's actions were announced in Navy Communique 369 on May 7, simultaneously with the award

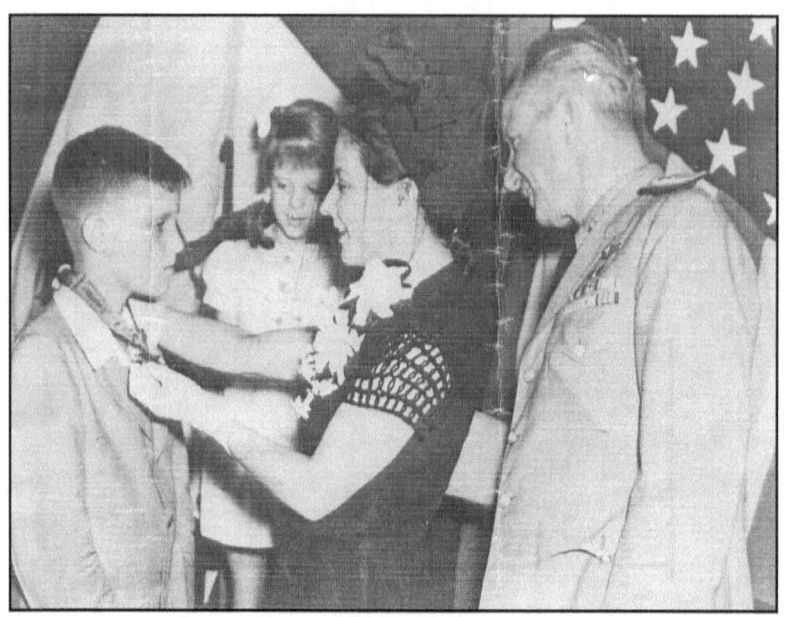

Gilmore's wife transferring her late husband's Congressional medal to their son, Howard Jr.

of the Medal of Honor by President Roosevelt. (This had been recommended by the task force commander, James Fife, and approved by Ernest King and Secretary Knox.)[41] Newspapers across the country carried the story, accompanied by various details. The New Orleans *Times-Picayune* reported "Navy men in Washington who had known Commander Gilmore said he was a quiet, mild-mannered man, likely to reach quick decisions without becoming upset by the circumstances which forced them."[42] The New York *Herald-Tribune* quoted Hilda: "My husband was a wonderful man, he loved the Navy and always want-ed submarine duty. It was his life, and he said that if he had to go he wanted it to be just the way it was—on the bridge."[43] In a July 13 ceremony in New Orleans, Gilmore's Medal of Honor and Purple Heart were given to his wife, son Howard Junior, aged 10, and daughter Vernon Jeanne (Darlene), 6.

Gilmore's family in their New Orleans home

Gilmore was commemorated during the war by the commissioning of *Howard W. Gilmore* (AS-16), in September of the year he died. "Happy Howie" served until 1980 and was scrapped in 2006.[44] Gilmore Hall houses the Enlisted Nuclear Power School at the Submarine School in New London, and a Navy housing project in San Diego is named Gilmore Terrace.[45]

Gilmore has two memorials at the Academy: a plaque outside his old room in Bancroft Hall, 7046[46], and one in Memorial Hall listing all the Medal of Honor alumni of the Naval Academy. In addition, his name is in the roll of honor under the Don't Give Up the Ship flag, listing those who gave their lives in the defense of our country. A display of USNA Medal of Honor recipients stands in the Visitors' Center, and there's a Gilmore Field across the Severn at Naval Support Activity Annapolis. Gilmore is also featured in the USNA Museum as a submarine hero of World War II.[47] His memory lives in literature as well: In Herman Wouk's *War and Remembrance*, an acknowledgment in the Epilogue states the death of the fictional Carter Aster was based on that of Gilmore.

What happened to his family reads, in the words of one researcher, like "a Greek tragedy." His son Howard Jr. was sickly and died young. His daughter Darlene dropped dead in the family home young also. Hilda lived on for several years afterward. All are buried in Hilda's family crypt in St

Gilmore's grave marker, courtesy of Ward Calhoun

Louis Cemetery # 3, New Orleans.[48]

Schade retired in 1971after a long career full of honors, and died in 2003 at the age of 91.

Perhaps the last person living who remembered Howard Gilmore was long-retired Bill "Radar" Hagendorn. Hagendorn served with Gilmore aboard *Growler*, but left the boat to have his foot fungus treated six weeks before she was lost in 1944. "He wasn't a very lucky man," Hagendorn observed from Tecumcula, California, where he tended a homemade graveyard and park on the slope back of his tract home, with concrete markers memorializing each of the fifty-two U.S. subs lost in WWII. "When the boat was launched, the champagne didn't break. It fell down in the dry dock and the electrician picked it up and threw it like a football. But he said, 'I don't want to hear anything more about this lucky and unlucky business.'

"He was a real guy. He wasn't stuffy. A right down-to-earth man. One time his wife came aboard. This will give you an inkling of what kind of man he was—she asked for a cup of coffee. The cup was all full of grease. And she said, "This is all full of grease.' And he said, 'What do you expect, it's free.'

"He was an enlisted man. Did you know that? He went in the Navy and took a test and became a student at Annapolis. But he still got along with everybody."

"There's not many of us left now," Hagendorn says, his voice almost vanishing. "Not many . . . we're just fading away."[49]

Hagendorn died a month after saying this, joining Howard Gilmore, the skipper who gave his life to save his men, and all his shipmates of the Submarine Service, on eternal patrol.

18
VICTOR "BRUTE" KRULAK, CLASS OF 1934: THE VARIETIES OF TRUTH

On the first day of Plebe Summer, 1930, a barely-over-five-foot seventeen-year-old with a receding chin stood at rigid attention on the yellow bricks of Tecumseh Court. Peering down, a distinctly unimpressed upperclassman sneered, "Well... Brute."

That shrimpy plebe became one of the most storied Marines in history. He helped midwife the landing craft that brought victory in the Pacific and Normandy; fought the Japanese in jungle battles; saved the Corps from a vengeful Army; helped develop vertical assault; spoke unwelcome truths to angry presidents. But he can also be seen as a master of denial, subterfuge, and embellishment. Victor "Brute" Krulak, three-star general, USNA '34, lived a self-crafted legend, the way he wanted to be perceived ... even when the facts contradicted it.

Was this admirable? Reprehensible?

Or should we rather ask: what truths are worth telling, in a time when those truths lead to prejudice or hatred; and what is it honorable to conceal, to accomplish a larger and perhaps nobler end?

Victor Harold Krulak was born in Denver, Colorado, in 1913, the only child of Meyer (later changed to Morris)

Krulak and Bessie Zelinsky (later Zall) Krulak. Meyer had immigrated from Russia in 1889; Bessie's parents came the same year. Victor later told author Neil Sheehan that his father had been a successful Denver gold-mine manager who had lost his fortune in the Crash.[1] In reality, Meyer had worked as a pawnbroker and watchmaker in Bessie's family pawn shop, until they moved to Cheyenne, Wyoming in 1919. There, Morris managed a clothing store.[2]

In those years, Cheyenne was a frontier town. The Jewish haberdasher's son learned to ride on an old horse his father bought from a nearby Army post. Victor might have had a Tom Sawyerish upbringing, like Philo McGiffin and Richmond Hobson. But his differed in an essential way. He was a member of a suspect and hated minority: the Foreign Jew.

It's hard now for us to recall the America of the 1920s, beyond the cliches of flappers and bathtub gin. But that period saw a surge in anti-Semitic feeling. Jews had never been popular in the United States, though a few individuals had been accepted for their talents. During the Civil War, U.S. Grant had gone so far as to expel all Jews from Kentucky, Tennessee, and Mississippi.[3] In World War One, "Jews were targeted by antisemites as "slackers" and "war-profiteers" responsible for many of the ills of the country. For example, a U.S. Army manual published for war recruits stated that, "The foreign born, and especially Jews, are more apt to malinger than the native-born.'"[4]

In the twenties, hatred of Jews merged with the mistrust of radicals. "Antisemitic traditions had existed in this country for centuries. Nevertheless, the antagonism toward Jews increased alarmingly during the postwar decade. Pseudo-Scientific racist thinking ... contributed to this animosity. So, too, did the 1917 Bolshevik revolution in Russia ... Jews were also attacked for allegedly leading the charge for a domestic socialist revolution that existed only in the minds of the most anxious Protestants and

Catholics."[5] Henry Ford's *Dearborn Independent* ran virulently anti-Semitic articles for ninety-one weeks straight starting in 1920, and his allegations were picked up by rural newspapers throughout the land.[6]

Hatred wasn't confined to print. The lynching of Leo Frank in Atlanta in 1915[7] was an extreme, but not isolated, instance of violence. Charges of "ritual murders" by Jews surfaced in Clayton, Pennsylvania,[8] and the Ku Klux Klan promoted its version of "True Americanism" with floggings, evictions, and cross-burnings.[9]

Victor Krulak's parents weren't religiously observant. Nonetheless, it's impossible to imagine him growing up unconscious of how the larger world regarded his "race."

Krulak's biographer, Robert Coram, has dug more deeply than any other researcher into these early days in his authoritative *Brute: The Life of Victor Krulak*, U.S. Marine. Charles Krulak calls it, "The best book about my father . . . A true historian, Robert wrote about the "good, bad, and ugly" of my Dad's life."[10] Coram speculates that Krulak was attracted to the military because it offered a college degree, demonstrated patriotism, and represented entree into the upper reaches of a society from which Jews, in those days, were excluded.[11]

But it was a Faustian bargain. In order to be accepted, a Jew could not remain what he was. He would have to become something else . . . as Albert Michelson, Class of 1873, and Hyman Rickover, 1922, had before him.[12]

But suddenly, the unexpected. A few months before he was due to take the entrance examination for Annapolis, the sixteen-year-old Krulak eloped with the fifteen-year-old daughter of an Army major. He signed the marriage register with the name "Donald Merrell." Their union was annulled, and the girl's family left town.[13]

Victor failed his first entrance exam. But Morris used his influence with a local congressman to renominate him, after the boy had boned up at the Bobby Werntz

Preparatory School (a sort of NAPS-for-civilians). "Many candidates came to Annapolis to prepare for the test at schools such as Wilmer and Chew's and Bobby Werntz's."[14] He entered USNA in June, 1930.

Krulak never said much about his time at the Academy.[15] At USNA at that time, "nativist, elitist snobbery and some incidents of anti-Jewish behavior" were not unknown.[16,17] But his assimilation (or perhaps protective camouflage might be a better term) was probably aided by the fact that until about 1938, even practicing Jews were required to attend the Academy's Protestant services.[18]

His main sport was crew, but he didn't row. A photo of him with the team shows him standing a full head smaller than the rest. That made him a natural for coxswain. "His name was inscribed on the Crenshaw memorial cup in 1932 and 1933. Each year, the name of the coxswain of the eight-oared crew winning the greatest number of races for that year is inscribed on the cup. He was elected captain of the crew squad for 1933–34."[19] This apparently was the height of his leadership achievements, as his highest midshipman rank was Second Petty Officer, Eighth Company.

His most notable conduct infraction came his Firstie year, when he was awarded fifty demerits and 14 days' punishment aboard the station ship *Reina Mercedes* for a Class 1 offense: illegally selling Beat Army stickers.[20] His best grades were in English and history; his worst, in aptitude.[21] His summer cruises in 1931 and 1933 were aboard *Wyoming*. During the summer of 1932 he had aviation instruction at USNA,[22] twenty-five hours in "Yellow Peril" floatplanes at the aviation facility across the Severn (now used for boat storage for Naval Station, Annapolis).[23]

But an acquaintance he made then would help his later career. Through his classmate and friend John Victor Smith, Krulak met Smith's father, then-Lt Colonel Holland M. Smith, a hero of Belleau Wood and probably the reason Krulak decided for the Marine Corps.[24]

The little mid also met Amy Chandler, petite daughter of his electrical engineering professor, at a USNA dance. The Chandlers were "an old, genteel East Coast family, Episcopalians all."[25] He would marry her two years after he graduated.

Krulak graduated 155th out of 463 in his class,[26] but another hurdle lay ahead. Due to the Depression, the Navy could not afford to commission all the graduates. Krulak, since he hadn't really grown since plebehood, failed the height and weight requirements. (Actually he was about the same height and weight as another Marine hero, Smedley Butler.[27]) Krulak liked to tell a story about having a classmate hit him on the head with a board, so the resulting bump would elevate him to acceptable standards; actually, he required a waiver, which Holland Smith helped him get.[28]

The new second lieutenant's first assignment after The Basic School was with the Marine Det aboard *Arizona*. His official record shows this followed by an assignment at USNA, but the Annual Registers for the 1935 and 1936 class years do not list him as an instructor or postgraduate student.[29] He was actually training for the Olympics. Krulak had impressed the crew coach at that time, the renowned Buck Walsh, who requested him as the coxswain of the 1936 Varsity crew that would compete for the privilege of representing the U.S. later that year. Unfortunately Navy got off to a shaky start in the final intercollegiate race and finished third. After that, Krulak was detached from the Academy and sent back to the Marines.[30] In 1937, he sailed for Shanghai, where he served with the 4th Marines for two years as a company commander.[31]

Shortly thereafter, Krulak, as an assistant intel officer, finagled permission from the Japanese to observe an amphibious assault against the Chinese at the mouth of the Yangtze. Krulak, with George Phelan, USNA '25, went in among the gunfire support destroyers and transports in a Navy tug, flying a huge American flag. "And there we saw, in action, exactly what the Marines had been looking for—sturdy, ramp-bow-type boats capable of transporting heavy vehicles and depositing them directly on the beaches."[32] Krulak sent the Bureau of Ships sketches, photos, and calculations, but his report landed in the dead files, noted as the work of "some nut out in China."[33]

In 1939 Krulak and Amy left China for Quantico, where he attended the Junior School (later the Expeditionary Warfare School).[34] Here he re-encountered Holland Smith, who now commanded the First Marine Brigade. Since Smith had been involved in developing amphibious doctrine since at least 1920,[35] Krulak built a balsa model of what he thought a bow-ramped landing craft should look like.[36] On Brute's graduation, Smith asked for Krulak as an aide, and assigned him to develop amphibious doctrine and landing craft to execute it.[37]

Both were badly needed. As late as 1941 the Navy had no craft suitable for landings. But Andrew Higgins, a New Orleans boatbuilder, had designed a novel hull for Mississippi rum-runners during Prohibition.[38] He'd offered the "Eureka" to the Navy since 1927, and had been working with the Marines since 1934.[39] In 1941 Smith sent Krulak to help Higgins fine-tune his design. This is the point at which, according to Coram, Krulak contributed the idea of the bow ramp.[40] (Smith, however, writes in his memoir *Coral and Sand* that Brigadier General Emile P. Moses, of the Marine Equipment Board, "worked out the idea with Higgins in New Orleans."[41]) After a head-to-head contest with a Navy design, Higgins's was chosen. It became the famous LCP(R)s, LCVPs and LCMs.[42] Higgins, Dwight D. Eisenhower said years later, "won the war for us."[43]

Meanwhile, the eccentric Donald Roebling had been developing a tracked rescue vehicle after the Great Lake Okeechobee Hurricane of 1928, in which over one thousand eight hundred Floridians drowned.[44] Roebling

Seaplane being launched at NAF Annapolis, undated, from Nimitz Library

had been working on his machines since 1932, and with the Marines since 1937. Once again, Krulak got on the field late in the game, in February, 1941, when Smith put him in charge of the test program of the "Alligator" in Puerto Rico.[45] Krulak, driving, stranded Ernest J. King (USNA '09) on a coral reef, forcing the Commander, Atlantic Fleet, to wade back to shore in his dress whites.[46] Krulak did compile a list of improvements and suggestions, which became part of the requirements for the LVT (1).[47]

When World War Two began, Krulak volunteered for training for a new sort of unit: Marine paratroopers. He took over the 2nd Parachute Battalion, First Marine Parachute Regiment, First Marine Amphibious Corps (IMAC) in New Caledonia. Though his men were initially unimpressed, Krulak won them over with a combination of toughness and concern. Injured badly during a training jump, Brute still led a twenty-mile forced march. He learned the names of all six hundred of his Marines, and knew them by sight.[48]

Krulak in official Marine Corps photo

Krulak in official Marine Corps photo in Vietnam

From August to December of 1942, the Marines had stonewalled the seemingly inexorable Japanese advance at Guadalcanal. Krulak's battalion landed on Vella Lavella in October 1943. Weeks later, the newly-promoted lieutenant colonel reported to Major General Archer Vandegrift's headquarters to discuss a special mission.[49]

The Paramarines would land at Choiseul, an important base for Imperial barge traffic. "There they (Krulak's men) would conduct diversionary raids on Japanese fortifications on the northwest part of the island and make the Japanese commanders think there were more Allied troops than there actually were."[50] If "Operation Blissful" worked, the enemy would reinforce Choiseul and weaken Bougainville, where the Third Marines were scheduled to make a major landing. To supplement their M1941 Johnson rifles, acquired from a shipment originally intended for the Dutch East Indies,[51] Krulak would get new Johnson light machine guns, mortars, bazookas, and a new 4.5-inch rocket . . . but there were six thousand Japanese on the island, maybe more.[52] The six hundred Marines would be massively outnumbered, and the enemy had air superiority. Krulak was assigned an Australian guide, Cardon Seton. Seton was six foot two, a "formidable presence", who had coastwatched on the island for over a year under the noses of the Japanese.[53]

The Second Battalion landed from LCP(R)s at Voza, on Choiseul, on the night of October 28. Guided by Seton, and assisted by Choiseulians, they moved their gear and ammo a half mile uphill to a jungle camp. That night Krulak laid plans for an attack on the garrison at Sangigai, to the east. Simultaneously, another patrol would go west and attack Moli Point. The battalion's radio operator began simulating the traffic of a 20,000-man landing.[54] Adm. William Halsey (USNA '04) backed this up with a press release from his headquarters, a deliberate misinformation of the U.S. press.[55] Krulak told his officers, "We need to

Krulak in 1943

make noise ... to make the Japanese believe that an Allied invasion has begun in earnest. To do that we need to hit them with a lot of small units over a broad front all at once. We need to look big. With that in mind, don't let your patrols get locked into a fight. Hit and run. Kill as many as you can and make a lot of noise, but get out."[56]

Immensely outnumbered, in an unfamiliar, malarial, snake- and scorpion infested jungle, with untested weapons, in his first combat mission after a career in staff billets, Brute had to take on the most ferocious enemy the Marine Corps had ever encountered: the Imperial Japanese Marines.[57] He wanted to see the enemy for himself. Krulak set out before dawn October 29 toward Sangigai with 19 marines and five native guides. Eventually he ran into ten Japanese unloading a landing barge. Deploying his men, he unslung his carbine and whispered, "Pick out individual targets. When I give the signal, kill them all."[58] They killed three Japanese, set C-2 explosives on the barge, and left it burning behind them.

Back at camp, Krulak sketched a complicated plan of attack. Each element would hit the enemy simultaneously. But that afternoon a Japanese patrol struck one of his outposts. They knew where the Marines were.

The next morning, Krulak and the Sangigai raiders set out. They trekked east, crossing the Vagara River. Then Krulak split his small force once again. F Company would move inland, into the jungle, and attack Sangigai from the rear. E Company would attack along the coast. Both assaults would occur precisely at 1400. Krulak led F Company on a five-mile trek that proved more grueling than he expected. His men noticed he was carrying his own pack and ammunition, though he was the smallest. He led from the front, urging them to keep up the pace, but they fell behind schedule.

E company ran into about three hundred troops on the beach, formed a skirmish line, and began firing as the enemy faded into the jungle. But F Company was still far from the action. When they finally made contact, the Japanese had manned a pillbox atop a hill the Marines needed. Instead of executing an ambush, many had the impression they'd walked into one, complete with preset firing lanes and snipers.[59] Krulak was prone, firing his carbine at the enemy, when a bullet blew out the back of his left arm.[60] But with his officers and sergeants, he kept moving forward, while siting his light machine guns and mortars to return fire.

A lull in the firing allowed both sides to yell insults at each other, such as "The Emperor is a queer" and "Babe Ruth is a son of a bitch." Krulak took advantage of the pause to emplace his machine guns. He wanted to get behind the Japanese, use his flanks and Easy Company to contain them, and wipe out as many as he could.[61]

The lull gave way to blood-curdling shrieks as the Imperial Marines worked themselves up for a banzai charge. "Fix bayonets," Krulak ordered. With only about twenty to thirty yards between the two lines, fighting would be hand to hand.

Screaming, the Japanese charged downhill. They met a storm of grenades, and full automatic fire from a Johnson A-4, with Krulak, his face a mask of blood from shrapnel, feeding ammo.[62] The attack faltered, and the survivors retreated into the jungle.[63] The U.S. Marines were left in possession, but holding ground wasn't their mission.

After burying their dead with explosives, they began the retreat to Voza.[64]

The report Krulak filed after his return read:

> The SANGIGAI operation commenced with outpost activity in the vicinity of VAGARA VILLAGE at 1100, October 31st. The enemy outpost was driven into SANGIGAI by a force moving down the coast while another force moved inland through the mountains and swung west to attack the Japanese rear. The main force was struck at about 1400. It quickly abandoned SANIGIGAI almost to a man, withdrawing into the mountains direct in the face of the enveloping force. Contact was made by the enveloping force at 1420 when the Japs were struck from two sides. This phase of the action, which lasted for forty-five minutes, was a fire fight of the most vigorous sort. During its progress the Japs undertook two of their customary Banzai counter attacks, during which their losses were great, and the failure of which caused their defeat. About 40 ran in a most un-Samurai fashion and escaped to the southward. Seventy-two dead Jap marines were counted. Our casualties were six killed, one missing and 12 wounded.....[65]

Company E had found a trove of intel in Sangigai, including charts showing mined areas and cleared lanes off Bougainville. This was so vital to the upcoming landing that the Navy sent in a PBY to pick it up, and to take out the wounded.[66]

In the following days, Krulak used his Higgins boats, hidden on a small island offshore, to insert and withdraw his forces. A second raid aimed at Choiseul Bay, and led by Krulak's exec, got bogged down and surrounded. Krulak, back at Voza, asked for PT boats from Vella Lavella to extract them. One was captained by a young lieutenant named John F. Kennedy.

"Kennedy had only one-third of a tank of fuel in PT-59, which was enough to get them to the Warrior River but not enough to get them back. The two officers decided that the PT-59 and PT-236, already fueled, would leave immediately. When Kennedy's boat ran out of fuel, the other boat would tow it. "Wind 'er up," Kennedy told his crew as they left the cove and headed out. The two PT boats reached the Choiseul coast and began looking for a landing craft to help guide them . . . At around 6 p.m. Kennedy spotted the boat 300 yards off Voza. On board were Krulak and Keresey, who immediately transferred to PT-59."[67]

Or maybe not. Although Marines interviewed after the war remembered Krulak as being there,[68] and Brute bragged about meeting Kennedy there, including a story about promising him a bottle of Three Feathers whiskey,[69,70] Coram says Krulak wasn't, and that he recanted in an addendum sent to the JFK Library.[71] But the JFK Library says no such addendum exists.[72] At any rate, whether Brute was there personally or not, the PTs extracted the marines in a rainstorm, though Kennedy ran out of gas on the way back and had to be towed.

Firefights grew heavier as the Japanese began closing in on the camp at Voza. Meanwhile, their command pulled troops out of Bougainville to send to Choiseul—exactly as General Vandegrift had planned. D-Day at Bougainville went off smoothly, and although Krulak was scheduled to stay on Choiseul for another week, early on Nov. 3rd IMAC sent him the message, "Sizeable enemy force converging on Second Marine Parachute Battalion from both flanks." They "advised" him to withdraw that day. Krulak felt he could hold out, but his raids had achieved their desired effect.[73] That night, screened by PTs, again including Kennedy's, the Paramarines embarked on three LCIs, leaving a slew of booby traps for the Japanese.

PT-59 near Vella Lavella, from John F Kennedy Library

Krulak was awarded the Navy Cross and Purple Heart for Choiseul. While the Paramarines were merged with the newly formed 6th Marine Division, he underwent several surgeries at Bethesda. He was then assigned to Quantico, where he assumed two responsibilities: to improve the boats and amtracs, in preparation for further campaigns, and to help derail General Marshall's plan to abolish the Corps, as part of a postwar reorganization.

Late in 1944, Krulak went back to the Pacific as assistant chief of staff of the 6th Marine Division, in charge of training for the upcoming invasion of Okinawa. He pushed physical readiness hard, as he always had.[74] Okinawa was the largest naval operation in history, with over one thousand three hundred ships and more than a hundred thirty thousand troops. Brute was often in the front lines, advising and coordinating as the troops advanced. He earned the Legion of Merit with "V" device for valor there. But in August, A-bombs fell and the war was over.

After a spell with the Sixth Marines in China, disarming Japanese troops, Krulak went back to Quantico. "Unification" was in the air. For the Army, that meant transferring Marine air squadrons to the Air Force, and stripping the

Krulak, official Marine Corps photo

Corps of its land combat role, returning it to the shipboard police force it had been before the Civil War.[75,76] This was an era of savage budget cutting and dependence on the atomic deterrent, when a sitting Secretary of Defense could write, "We'll never have any more amphibious landings. That does away with the Marine Corps. And the Air Force can do anything the Navy can do nowadays, so that does away with the Navy."[77]

At Quantico, Vandegrift charged several officers—Jerry Thomas, Merritt Edson, Bill Twining, Krulak, and others—with putting a stop to the unification plan.[78] The "Little Men's Chowder and Marching Society," as they became known, managed to get their hands on secret Joint Chiefs position papers which, in their view, not only threatened the Corps, but dangerously weakened civilian control of the military. The testimony they drafted for

General Vandegrift to present before Congress (the famous "Bended Knee" speech) aborted the plan. Krulak followed up with a nineteen-minute PR film about the island-hopping campaign that's still an impressive justification for the amphibious mission of the Corps.[79]

The revised unification plan for 1947 "dropped the idea of the single military chief of staff,"[80] but still left the Corps legally unprotected. During the year that followed, Krulak, Twining, Edson, and other Marines continued to outmaneuver Eisenhower and Truman. The final National Security Act of 1947, "written largely by Krulak"[81] preserved Congressional control of the military and legally defined the Corps' mission.

"The Brute" moved on to helicopters, at the time flimsy, undependable gimmicks. Once again, he did not exactly originate a new idea. The concept of using helicopters in vertical assault came from a general board commissioned by Vandegrift after the Bikini Atoll H-Bomb tests. (As far back as 1909, second lieutenant Vandegrift had written a Parris Island thesis on "Aviation, the Cavalry of the Future." It was graded "unsatisfactory."[82]) "On 16 December, 1946 the special board submitted an advanced report to the Commandant recommending that parallel programs be initiated to develop a transport seaplane and a transport helicopter. The board further recommended that an experimental Marine helicopter squadron be organized to train pilots and mechanics and that the Marine Corps Schools develop a tentative doctrine for helicopter employment."[83] Krulak took charge of developing doctrine, and arranged the first experimental helicopter assault in history, from a carrier off Camp LeJeune, in 1948. He also wrote the Marine manual for helicopter operations, concepts the Army used fifteen years later.[84]

Krulak next commanded the 5th Regiment at San Diego. He scraped and scrounged to train the one thousand eight hundred men he had, out of a "war strength" of three thousand nine hundred.[85] When the Korean War broke out, Krulak, now operations officer at FMF (PAC) in Hawaii, put his career on the line. He answered the CNO's request for how soon and how much the Marines could respond by promising a reinforced battalion in 48 hours and a reinforced regiment with an attached air wing in five days ... in his commander's name, but in his absence (General Lem Shepherd was fishing in Wyoming and couldn't be reached).[86] Shepherd and Krulak rushed deployment of a six thousand five hundred-man Provisional Marine Brigade, which held the line at Pusan.[87] Then, accompanying Shepherd to see MacArthur in Tokyo, Krulak drafted the message requesting the First Marine Division for a landing at Inchon.[88]

Meanwhile, helicopters "... proved themselves almost immediately upon their arrival in Korea, at Pusan. They became the "secret weapon" of the Marine Brigade's command and staff. These aircraft became the lifeline that brought thousands of wounded Marines to hospitals only minutes after being wounded. The HRS-1's of HMR-161, after their arrival in the fall of 1951, validated all of the thought that the Marine Corps Schools at Quantico had put into the helicopter."[89] Krulak went back and forth between Hawaii and the battlefront, including a helo landing at Inchon with Shepherd just after the airfield there was liberated.[90] In late 1951 he became Chief of Staff of the 1st Marine Division.[91] He earned a second Legion of Merit with Combat "V", and was awarded the Air Medal for reconnaissance and other flights in Korea between August 1950 and July 1951.

In late '51, the First Division had settled into a defensive mountain line around a valley called the "Punchbowl,"[92] beginning a long battle of attrition with the Chinese, when IMR-161, the first Marine helicopter transport squadron, arrived in theater. Krulak zealously pushed their employment, first using the Sikorsky HRS-1s to carry supplies, then to move troops to take a hill (Operation Summit), then a nighttime troop lift (Operation Blackbird). The Marines capped this with Operation Bumblebee, the first heliborne combat assault—what would later be called "vertical envelopment" when the Army "invented" it in Vietnam.[93] That November Krulak returned to Washington for duty at Headquarters Marine Corps as Secretary of the General Staff.[94]

This might be a good place to address the Krulaks' family life. As Krulak's three sons were growing up, he held them to extremely high standards. He tended to be distant and demanding, rewarding shortcomings either with icy silence, a session with their father's belt (more common, and indeed widely expected, in that era compared to today), or a whack on the head with his Academy ring. As his eldest, Vic, says, "Sensitive issues didn't come up." Middle son, Bill's view: "You knew where you stood. He was consistent and fair, tough but he never raised his voice in anger." Chuck says, "He was a hard man but he instilled in us a strong value system and a sense of the importance of character." He also thawed slightly as the boys grew to adulthood. Through it all Amy held the family together during her husband's long absences and was a moderating presence when he was home.[95]

Krulak served as Chief of Staff, Fleet Marine Force, Pacific, and was promoted to brigadier general, assuming duties as Assistant Division Commander, 3d Marine Division, on Okinawa. In 1957, he became Director of the Marine Corps Educational Center, Quantico. He was promoted to major general in 1959 and assumed command of the Recruit Depot, San Diego.[96]

In 1960, however, an ex-Navy PT skipper was elected president. In the version Krulak told in the 1980s, he took the initiative, sending Kennedy the bottle of whiskey along with a note reminding him of their wartime meeting. Then, Krulak said, Kennedy invited him to the White House.

"They enjoyed a ceremonial drink of the Three Feathers while they reminisced,"[97] and Kennedy himself requested Krulak as his advisor. In a later interview, Krulak modified his account: "I was thinking of our being together on the island of Vella Lavella where his squadron was based and where my battalion was in fact based. Well, I was wounded and evacuated. I ended up in the United States and I never saw Kennedy to give him the whiskey. But I kept the bottle. Three Feathers it was, doesn't mean anything to you probably, but Three Feathers was just the most wretched rotgut in the world and one day I was going to give it to this fellow. You know, I sort of lost track of him, but I owed it to him, and one day I was going to give it to him. I kept it, and it followed me around in my chattels.

"Then he got to be President and I was ordered to Washington. He sent for me and we chatted about the Solomons. I didn't bring the bottle of whiskey with me because I was, you know, sort of on my mettle. I wasn't going to do anything like that the first moment. But on another occasion I brought him the whiskey. Of course, he didn't remember that I had promised him a bottle of whiskey, but he knew the circumstances, and he accepted it with great enthusiasm. He said his problem would be deciding when to drink it."[98]

Well . . . maybe. Yet a third version is that when Kennedy wanted to de-emphasize massive retaliation in favor of counterinsurgency, none of the Chiefs wanted any part of it. They suggested nominating someone to report directly to Robert McNamara, the Secretary of Defense. It was the Army that suggested Krulak, hoping he would fall on his sword.[99]

Regardless of how it happened, in February 1962, relinquishing his command in San Diego, Krulak assumed duty as Special Assistant for Counterinsurgency and Special Activities.[100] Author David Halberstam wrote, "The war in Vietnam was Krulak's business. He was its inspector-general in Pentagon heraldry, overseeing the conflict from Washington on a day-to-day basis for the Joint Chiefs and McNamara. . . ."[101] He also worked for Maxwell Taylor, even clearing important telegrams on his behalf when the Chairman could not be located, just as he had for Shepherd.[102]

Krulak traveled to Vietnam in the joint Krulak-Mendenhall Defense/State mission of 1962. In the aftermath of Ngo Dinh Diem's brutal suppression of the Buddhist uprising in August, which Diem had solemnly promised the U.S. Ambassador he would not do, the administration faced three choices: 1) continue with Diem down a path which looked like defeat; 2) acquiesce in a military coup; or 3) disengage from South Vietnam entirely.[103] All three presented significant downsides. Kennedy sent Krulak, along with the State Department's Joseph Mendenhall, to determine whether the Army of the Republic of Vietnam was trustworthy.[104] They toured the country independently, and on their return, gave "diametrically opposing"[105] reports. Krulak, who had flown up and down the country, said the South Vietnamese military were staunch patriots and firm allies. Mendenhall, who'd visited only Saigon, Hue City, and Qui Nhon, said the army was corrupt and couldn't be counted on. Kennedy asked in bewilderment, "Were you two fellows in the same country?" [106] A second mission, the McNamara-Taylor mission in September, resulted in the suspension of U.S. aid. "As might have been expected (although the record leaves ambiguous whether this was a conscious aim of the Administration), the Vietnamese generals interpreted the suspension as a green light to proceed with a coup."[107]

Krulak was pulled off Vietnam during the Cuban Missile Crisis in 1962. His role in this remains shadowy. Before and after Cuba, as SACSA, Krulak wrote influential classified papers, gave briefings, wrote the Joint Counterinsurgency Doctrine, and met one on one with McNamara and JFK many times.

Krulak did not advocate massive intervention. Starting with a classic "small wars" mindset, he understood that Vietnam was a counterinsurgency—COIN, in the acronym of the day—rather than the counter-NVA or counter-guerrilla war officers like William Westmoreland viewed it as. Krulak said, "Serving in the Joint Staff as the focal point in counterinsurgency operations and training . . . I learned something of the complex nature of the conflict there. The problem of seeking out and destroying guerrillas was easy enough to comprehend, but winning the loyalty of the people, why it was so important and how to do it, took longer to understand. Several meetings with Sir Robert Thompson, who contributed so much to the British victory over the guerrillas in Malaya, established a set of basic counterinsurgency principles in my mind. Thompson said, "The peoples' trust is primary. It will come hard because they are fearful and suspicious. Protection is the most important thing you can bring them. After that comes health. And, after that, many things—land, prosperity, education, and privacy to name a few."

The question of whether Vietnam could have been won for democracy will never be answered. To some reporters, Krulak's insistence that victory was possible was a dangerous lie, signaling that his ambition had overcome his moral courage.[108] But the Republic of Vietnam might have maintained its independence longer if Krulak's recommendations had been followed.

That was not to be. In 1963 John F. Kennedy was assassinated, a Texan became president, and massive U.S. involvement became much more likely.[109] Krulak gained a third star as Commander, Fleet Marine Forces, Pacific. He trained his men hard, and soon they went to war, starting with the Da Nang landing in March 1965.

It's simplistic to say that the Army tried to win in Vietnam through body counts, attrition, and search and destroy, while the Marines tried to win through winning hearts and minds. The Marines definitely killed guerrillas, and the Army did try to win over the population. But their approaches were different enough that a gulf opened between Westmoreland and Krulak.[110] The Brute made fifty-four trips to Vietnam between 1962 and 1968,[111] and when he wasn't in-country, he met and walked through every planeload of Marine wounded that passed through Hawaii, shaking the hand of every man who was conscious.[112] Eventually all three of his and Amy's sons, two as combat Marines and one as a Navy chaplain, would serve in Vietnam.

As the war progressed, and more and more troops were poured in—and both casualties and domestic opposition increased—Krulak became convinced of two truths. The flow of supplies through Haiphong had to be stopped, and COIN was the right strategy, not attrition, which simple demographics made impossible. When Westmoreland did not concur, Krulak drafted a major paper in Hawaii, late in 1965. "He took a week to write it, drafting the sentences in pencil on a large pad of lined paper as he did all of his important letters and memoranda."[113] Krulak began urging his recommendations for a major change in strategy up the line.[114] Early in 1966, in Washington, he shared these views with McNamara and Averell Harriman, who felt the President ought to hear them. Krulak went to the Oval Office on August 1, for a famous interview with LBJ.[115]

One on one with Johnson for forty minutes,[116] Krulak, who was being considered for a fourth star as Commandant of the Corps, took the president to task. His message was plain: We would lose, unless major changes took place. To Coram, Krulak said: ". . . I told him if he did not change, he would lose the war and he would lose the next election."[117] In *First to Fight*, he wrote, "I spoke of the need for improvements in the quality of the South Vietnamese government and for acceleration in the training of South Vietnamese forces. But, most of all, I told him we faced a self-defeating attritional

Krulak in conversation with Lyndon B. Johnson in 1966

cycle involving engagements with large and increasingly sophisticated North Vietnamese units. We had to stop the flow of war materials to those forces ... Then I voiced the critical words, urging that we "mine the ports, destroy the Haiphong dock area." That was it. As soon as he heard me speak of mining and unrestrained bombing of the ports, Mr. Johnson got to his feet, put his arm around my shoulder, and propelled me firmly toward the door."[118]

It's worth emphasizing how rare this kind of event is in recent American history. As Halberstam wrote in his history of the Korean War: "Brave and otherwise independent men often became quite bureaucratic once they were members of the JCS. That reflected one of the great secrets of the military culture—how officers who had been so brave in battle, fearless when it mattered, could be so bland and cautious as they reached what was seemingly a career pinnacle. That had been true in Korea; it would be even truer in Vietnam. There were, it appeared, two very different kinds of courage in many military men—bravery in battle, and independence or bravery within the institution—and they did not often reside side by side."[119]

In Brute Krulak, they did. Others, equally brilliant, saw what he saw, and remained quiet for the sake of their careers.[120,121] His biographer writes, "In Vietnam, it was the job of active-duty senior officers to present unvarnished advice to the president, but those generals, in every branch of the service, were silent and servile . . . Krulak, as far as can be determined, was the only senior general who confronted LBJ."[122] This is even more astonishing when we recall how much Victor Krulak had denied or submerged to achieve success. He occasionally exaggerated, perhaps to make himself feel important, or to promote his legend. But when he was important, and in reach of the crowning honor of his career, he had the courage to speak the truth, and risk losing it.

And he did. Although he still had access to the president—including another one on one "off the record" meeting for 19 minutes on January 27, 1967[123]—LBJ selected another Marine to head the Corps. Krulak made one more visit to Vietnam, debarking from a helo under fire to visit his Marines during the battle of Khe Sanh, in Quang Tri province (he had fought against their being deployed there, feeling it was a NVA diversion, but lost to Westmoreland again.[124]) He retired in 1968, with 34 years of service.

"Brute" Krulak's decorations included the Navy Cross, the Distinguished Service Medal, the Legion of Merit with Combat "V" and two Gold Stars in lieu of second and third awards, the Bronze Star Medal, the Air Medal, the Purple Heart, the Presidential Unit Citation with three bronze stars indicative of second through fourth awards, the China Service Medal with one bronze star, the American Defense Service Medal with Base clasp, the American Campaign Medal, the Asiatic-Pacific Campaign Medal with three bronze stars, the World War II Victory Medal, the Navy Occupation Service Medal with Asia clasp, the National Defense Service Medal with one bronze star, the Korean Service Medal with four bronze stars, the Vietnam

Service Medal, the United Nations Service Medal, the Korean Order of Service Merit second class, the Republic of Vietnam National Order Medal third class, the Republic of Vietnam Gallantry Cross with Oak Leaf Cluster, two Korean Presidential Unit Citations, and the Republic of Vietnam Campaign Medal.[125] In 2007 the Secretary of Defense gave tribute to his memory, quoting Krulak's words about the "adaptability, initiative and improvisation (that) are the true fabric of obedience, the ultimate in soldierly conduct, going further than sheer heroism."[126]

Krulak stayed active after retirement, writing and serving on charitable boards. In 2004 he was named a USNA Distinguished Alumnus, and he served as Chairman of the Academy's Board of Visitors for four years.[127] His most enduring legacy, though, may be his book, *First to Fight*, published by the Naval Institute Press. In 2007, it was made required reading for all Marines.[128]

Victor's youngest son, Charles C. "Chuck" Krulak, USNA '64, assumed duties as the 31st Commandant of the Marine Corps on 1 July 1995.[129] His other two sons, Vic and Bill, also served honorably, and both became Episcopal priests. Amy died in 2004, and Victor Krulak in 2008, at the age of 95. He and Amy are buried in San Diego.

11

PAUL SHULMAN, CLASS OF 1945: "IF YOU CAN SINK THEM, SHOOT"

An admiral at age 26? First Chief of Naval Operations of a new navy? Few careers can match the rocketlike ascent of a quiet, introspective, "by the book" graduate of the Class of 1945.

Graduated with a war-accelerated class, targeted by kamikazes, Paul Shulman pivoted from being a junior engineering officer to running clandestine operations for a stateless army ... organizing a fleet from scrapyards ... directing a squadron in battle ... and serving as a fledgling navy's first Commander in Chief. After distinguishing himself in wartime, he was disinherited by Israel's military establishment. But he followed up with a long and fruitful career in engineering and construction, giving back loyally to the new state he had helped found and defend.

This is the story of Paul Nachman Shulman, '45. Though perhaps we should call him by his Hebrew name: Aluf (Admiral) Shaul Ben-Zvi.

Shulman was born in the Bronx in 1922, and grew up in Far Rockaway, Manhattan, and Connecticut. His father Herman was from Romania. His mother, Rebecca Bildner, was from Austria. Herman became a wealthy and influential lawyer, specializing in corporation law. As the family income grew, the teenaged Paul and his older brother, Mark, enjoyed a large apartment on Manhattan's posh Upper West Side overlooking Central Park. The boys enjoyed vacations in high-toned Fairfield County, Connecticut, at a summer house with tennis courts and a swimming pool.[1] Herman was as active in the Zionist Association of America as Rebecca was in Hadassah.[2] David Ben-Gurion, Golda Meir, and Abba Eban were frequent guests in their home. In 1944, Herman participated in the Dumbarton Oaks conference that helped turn the Moscow Declaration of 1943 into the charter of the United Nations.[3]

After prep school at Cheshire Academy in New Haven, then the only private school that openly accepted Jewish students,[4] young Shulman went to the University of Virginia. He joined an experimental OCS prep program, but in 1941 received a congressional appointment to the Naval Academy.[5]

Just after Pearl Harbor, Paul went home on Christmas leave to find Ben-Gurion there as a weekend guest. As Shulman told it later, the Zionist leader said "....he was glad to see that Jewish boys were studying to be naval officers; that we would need them some day in the Israeli navy. I said to him: 'I hope he had a fine navy, because I was planning to spend my life in the regular American navy.' Eight years later when I reported to Israel, to join the Israeli navy, he reminded me."[6]

Shulman didn't stand out from the crowd at USNA. He seemed more interested in professional matters than sports or academics. Though not overly observant, he may well have formed part of the Jewish Church Party, which in those days fell in early Sunday (yes, Sunday) inside Gate 1 and marched in ranks two blocks up East Street to meet at the Kneseth Israel synagogue. Kneseth Israel was an Orthodox congregation, but the non-Sabbath event the mids attended seem to have been less a worship service than a speakers' forum, and perhaps a chance to rub elbows with girls from out in town.[7]

Shulman's *Lucky Bag* entry reads, "This wise old man from Connecticut, the author, editor and publisher of Shulman's Fighting Ships,[8] showed more than his share of professional interest and knowledge. As Professional Editor of The Log and a contributor to Reef Points, Paul was always a reliable source of information about the Navy, particularly anything that concerned his favorite branch, submarines. Quiet and a hard worker, "Hap" always managed to keep ahead of the academic departments and maintain his outside interests. As an ex-N.R.O.T.C. man, "Navy" experienced no shock in adapting himself to the system, and as a stern disciplinarian he was a constant aid to others who were just beginning their careers in the Navy."[9] His sports were soccer and batt tennis. His best standing was in EH&G, his worst in math, but contrary to what the *Lucky Bag* implies, he hovered near the bottom of his class.[10] Of the 914 cadets commissioned with the Class of 1945, Shulman was ranked 835.[11]

The Class of 1945 (which included 32 Jews)[12] walked early, in 1944, and immediately went to the Fleet. Shulman wanted submarine duty, but the Navy needed aviators more. He was sent to NAS Jacksonville for flight training, but was not accepted, perhaps because of his eyesight.[13] He finally went to war assigned to a Fletcher-class destroyer, USS *Hunt* (DD-674).

In the oral history Shulman recorded the year before he died, he's reticent about his U.S. service, but the facts can be retrieved. Launched in 1943, *Hunt* had screened Marc Mitscher's Fast Carrier Task Force 58 in operations against Kwajalein, Roi, Namur, Truk, Jaluit, Palau, and Pelelieu.[14] Shulman joined as the assistant engineering officer (electrical),[15] and went through the terrifying Typhoon Cobra (December 1944) aboard her, when three other destroyers capsized and sank. In late 1944, *Hunt* participated in the strikes against Formosa and Japanese-held areas in the Philippines.[16] All told, Shulman would spend about a year and a half of combat service in the Pacific.[17] He was remembered as a quiet officer who insisted on proper Navy protocol. But he also showed bravery; as quarterdeck OOD, he once dove over the side in an unsuccessful attempt to rescue a crewman who had fallen off the brow when crossing within the nest.[18]

In later life, he recalled two battle experiences most vividly.

March 19, 1945 dawned with the sea calm and twelve knots of wind from the east. The sky was overcast with occasional breaks and excellent horizontal visibility. *Hunt* was screening USS *Franklin* as the carrier prepared to launch the first naval air strikes on the southern home island of Kyushu. The night before, the task force had gone to general quarters twelve times.[19]

A little after 0700, a Yokosuka-based D4Y "Judy" dropped two 500-pound bombs on *Franklin.* The carrier's decks were crowded with fueled aircraft loaded with bombs and rockets. In minutes the carrier was a flaming tomb, without power and dead in the water. Hundreds of her sailors were blown off the decks, or jumped overboard to avoid the flames. *Hunt* moved in toward the burning flattop to pick them up. Shulman supervised hauling some of the survivors on board.[20] "After rescuing 429 survivors, (*Hunt*) joined three other destroyers in a clockwise patrol around the stricken ship which had gone dead in the water within 50 miles of the Japanese Coast. Cruiser Pittsburgh (CA-72) took the ship in tow and, after an epic struggle, managed to get her to Ulithi 24 March. Hunt put the survivors ashore and sped to the Ryukyus 5 April to support troops who were struggling to take Okinawa."[21]

His second close shave came when *Hunt* was assigned as a radar picket off Okinawa, to detect attacks on the task forces. On April 14, it was her turn to be attacked. The kamikaze, although hit by *Hunt's* guns during its approach, still struck at deck level, shearing off the mainmast and leaving one wing in the forward stack. Its fuselage continued on into the water, while *Hunt's* crew fought the fires it left behind.[22] Shulman said, "We were very fortunate. Three men jumped overboard in fright. But we picked them all up ... we lost no one. The ship next to ours lost seventy-five."[23] He also noted, "One thing you learn ... When you're in the middle of combat you don't have time to be frightened. You can be frightened before or after. When it's happening there's no time to be frightened."[24]

Hunt stayed on station, alternating antiaircraft guard for the carriers with periods of more picket duty. When she finally left the Ryukyus, her crew had been to general quarters fifty-four times. *Hunt* finally sailed for the West Coast in June 1945 for an overhaul before the planned invasion of Japan.

Shulman's life changed while *Hunt* was in the States. First, he married Rose Saxl in San Francisco. Shortly thereafter, his father died. In August, two atomic bombs made his ship's return to Japan unnecessary, and she was decommissioned at San Diego.[25] Shulman was reassigned to USS *Massey*, (DD-778), which was being transferred to the Atlantic Fleet as a training vessel, homeported at Naval Base Norfolk.[26]

Shulman's path from the U.S. Navy to a shadowy cloak-and-dagger career was traced by retired Navy Chief Journalist John Wandres in *The Ablest Navigator: Lieutenant Paul N. Shulman, US, Israel's Volunteer Admiral*, published in 2010 by the Naval Institute Press. Much of what follows about that part of Shulman's life is from Wandres's excellent book—heartily recommended for greater detail about the man and his times—and his comments on early drafts of this piece. Murray Greenfield's *The Jews' Secret Fleet* was also useful in reconstructing Shulman's undercover activities.

Aboard *Massey*, Shulman read about the Royal Navy blockade of refugee vessels taking Holocaust survivors to what was then Palestine. In August 1946, he submitted his resignation. It was accepted, and he left the ship, and the Navy, in January 1947.

Shulman's first outside job was in import-export, but "import-export" shaded into two shadowy agencies, closely watched by British Intelligence and the FBI, which sourced weapons, funding, and personnel for secret militias of a nascent foreign state. In March 1947 a position was arranged for Paul with a company based in New York. In reality the job provided cover for Shulman's desire to help David Ben-Gurion. Bothered by reports of Great Britain's decrees to stifle Jewish refugees from emigrating to Palestine, Shulman had volunteered to help the Mossad le Aliyah Bet. "Mossad" is Hebrew for Department; "Aliyah Bet" means Immigration B, in the sense of "Plan B." If Jewish refugees could not legally immigrate (make Aliyah) to the homeland, they would have to get there by other means.[27]

Through his family, Shulman knew the men and women in charge of this effort, including Meir, Ben-Gurion, and Shlomo Rabinovitch. This period is still shrouded in mystery—it would take an espionage novelist like Alan Furst to really give the flavor of it—but Shulman, having naval expertise, was quickly passed along to one Akiva "Kieve" Skidell, who was recruiting war veterans for the Haganah.

A little background may be in order here. The Mossad le Aliyah Bet and Haganah were separate entities. Mossad's mission was to transport refugees to Palestine. Haganah's purpose was to develop and train a land-based fighting force. The two were not always hand in glove, and their competitiveness would affect Shulman's later career.[28]

Shulman's New York connections led him to one Ze'ev "Danny" Schind. Operating out of a penthouse office in Hotel 14 in Manhattan, Schind was in charge of acquiring ships to be converted to transport refugees. Assuming their office was bugged by the FBI, Schind would sometimes meet contacts in the hotel's basement—the Copacabana Supper Club. The joke in the penthouse was that the headquarters was known as 'Club CopaHaganah.'[29,30]

Shulman probably helped purchase the icebreaker ex-USCGS *Northwind*, the former Normandy comm ship *President Warfield* (bought for $8000)[31], and was soon negotiating for two ex-Central American banana transport ships,[32] *Pan Crescent* and *Pan York*. He later wrote, "There was a tremendous surplus of vessels, many of which were heading for the scrap yards. Vessels could be bought by weight for a few dollars on the ton. The Pan ships, for example, were bought for $125,000 each."[33] Converted into Spartan transports, these ships could take hundreds of thousands of displaced Jews in European refugee camps to the new land.

If they could break the British blockade, that is.

"On March 17, 1947, two weeks before his twenty-fifth birthday, Paul Shulman became president of 'F. B. Shipping' and owner of a pair of 360-foot-long refrigerator ships. Shulman said, "The B. Stood for Britain. And the F, you can use your imagination."[34] Quickly converted, both Pans left for the Med that summer under Panamanian flags.

President Warfield sailed into history as *Exodus 1947*. It left France with over 4,500 displaced persons, mostly Holocaust survivors. But before it could reach Palestine, British destroyers intercepted it. Its American second mate and two passengers were killed and dozens wounded. In Haifa the passengers were transferred to three British deportation ships. They refused to disembark in France, went on a hunger strike, and were finally transferred to camps in West Germany.[35] The embarrassment for Britain helped swing worldwide sympathy toward the Jews.[36]

Meanwhile, *Pan Crescent* had broken down and was in a shipyard in Venice. Dispatched to get her repaired, Shulman also found himself assigned as the "naval aide" to Aliyah Bet's Italian section chief. Arrested and interrogated by the Italian police,[37] he had barely been released and begun work when an underwater explosion sank the ship at the yard. The British blamed the Arabs; Aliyah Bet blamed the British. Shulman himself thought the ship a total loss, but pitched in when others disagreed. It was eventually refloated, repaired, and slipped out to sea under the nose of a British battlecruiser.[38]

Years later, Shulman remembered the trouble he had getting out of Venice that night with his crew. "The last thing he needed was to be publicly identified as a former U.S. naval officer. But the captain of a U.S. Naval ship who knew Shulman all too well was standing in his hotel lobby the night they were leaving. Years earlier, while on shore patrol in a South American port, Shulman had run the captain out of a brothel. To avoid contact with the officer, Shulman hired a prostitute and told her to tell the man that she was a 'gift from an old friend.' When the captain went off with the prostitute, Shulman and the Haganah men left without being seen."[39]

In March of 1948 Shulman went to Israel, appointed as "chief of staff for naval training." He was supposed to set up a naval academy in Haifa, but had to do so in the shadows of British occupation and without funds or a staff.

On May 14, 1948, David Ben-Gurion, as head of the Jewish Agency for Palestine, proclaimed the State of Israel. President Harry S Truman recognized the new nation on the same day.[40] The British High Commissioner left his keys on the table—literally. One day later, forces from Egypt, Syria, Jordan, and Iraq entered Palestine.[41] Shulman said, "But within two or three days I was called by Ben Gurion. He said, 'You are no longer the advisor. You are the first Commanding Officer.'"[42] Of the new Navy, that is.

Shulman had no staff, funds, equipment, or doctrine. He had to scrounge cast-off police uniforms, and Ben-Gurion

Exodus 1947, *formerly* President Warfield

A sleeping quarter on board Exodus 1947

gave him the arbitrary rank of Kvarnit—equivalent to a USN commander. The refugee carriers Mossad had purchased had been intercepted at sea by the British, disabled by their crews, and left to rust along the breakwater of Haifa Harbor. There wasn't even a Hebrew word for "navy"—The best equivalent was "Army of the Sea"—but he did have one operational ship—the former *Northland,* renamed *Eliat.*[43]

Shulman recruited instructors where he could, ending up with officers from four different navies,[44] and many of his students were from a covert force of naval commandos, the Palyam. There were women vounteers from abroad, too; they served as nurses in the Bat-Galim Base Hospital and in administrative positions in various shore bases and the headquarters in Stella-Maris on the Carmel in Haifa.[45]

In July he returned to America to recruit more ex-USN instructors, and also to purchase three surplus Canadian sub chasers, five hundred copies of the *Bluejacket's Manual,* fifty copies of *The Watch Officer's Guide,* and a telescope for the one-eyed Moshe Dayan.[46] Jonathan Leff, USNA '43, showed up in Haifa, and Shulman made him a gunnery instructor.[47] At some point during this chaotic, dangerous period Shulman was also dual-hatted as both a squadron commander and the overall leader of the Israeli Navy, corresponding to a blend of Admiral of the Fleet and CNO. He reported to the Prime Minister and Minister of Defense Ben-Gurion, but had to struggle in a welter of competing interests and lack of resources, including the new Israeli Defense Forces' separate command structure and the native Israelis' suspicion of the foreign volunteers, called "Machal" (literally, volunteers from abroad).[48]

Somehow Shulman managed to improvise. He pulled together aging, rusting ships that had lain abandoned for months or years, surplus subchasers, and patrol craft. He selected *Wedgwood,* the former Flower-class corvette *Beauharnois,*[49] remaindered by Canada, as the flagship.

Perhaps the most incredible moment of Shulman's career came that June, during the first truce period, when he got his improvised squadron—*Wedgwood, Eilath* and *Hatikvah*—underway to intercept an Egyptian force near Tel Aviv.[50] The Egyptian squadron was heavily armed; Shulman had only one 75-mm fieldpiece on *Wedgwood's* deck. In an amazing display of chutzpah and courage, he climbed down into a whaleboat with a loud hailer. Making up to the enemy flagship, *Princess Fawzia*, "He called up to the captain, and in English informed him that the Egyptian warships had illegally invaded the waters of the sovereign nation of Israel. Shulman instructed the Egyptian captain in no uncertain terms that it would be best if his ships took their departure."[51] This astonishing bluff actually worked; the hostile squadron departed!

His next engagement, though, was not to be as bloodless. On June 21, Ben-Gurion ordered him to intercept a former USN ship, *Altalena* (ex-LST-138), which had been bought at surplus by the Irgun Etzel (Jewish freedom fighters)[52] and loaded with arms and ammunition for what was identified as a rival army of the Haganah. Shulman accomplished the task, but not without loss of life aboard *Altalena*. However, there was only one army and one navy in the IDF thereafter.[53]

By September 1948 Shulman's desperate organizational efforts were paying off. Uniforms, antiaircraft guns, radar, and more ships were becoming available. Haganah fighting units, and the Israeli naval squadron and its crews, were finally getting the equipment and supplies they needed. Much of the materiel was available—at a price—on international arms surplus markets.[54] On October 19, Shulman put to sea a squadron of two corvettes and two patrol craft, shadowing another Egyptian reinforcement flotilla. But this one would not be an easy target. "The Emir Farouk . . . usually stayed within the protective range of coastal batteries. The small Israel Navy could not sink it with conventional methods."[55]

But the corvettes carried a secret weapon on their decks: four war-surplus MTM motorboats. Developed in Italy, these "crash boats" had destroyed the British heavy cruiser HMS *York* and a Norwegian tanker in 1941, and later supported Operation Barbarossa in the Black Sea.[56]

These near-suicide craft, fitted with heavy demolition charges in the bow, were manned by Shayetet-13, a naval commando unit hand-picked from among the special operations personnel who had served in the Palyam. Their commander was Yochai bin-Nun[57,58], and this early special operations unit include four American volunteers.[59] The Egyptian units—a minesweeper and the *Emir Farouk*, flagship of the Egyptian fleet—moved slowly toward Gaza to disembark troops.

But on October 20 a UN-brokered truce had been declared.

Both sides had broken previous truces, and Shulman, in tactical command aboard *Wedgwood*, was prepared to break this one. Some sources say that before that, though, he "pulled alongside the Egyptian ships and Shulman called out over a loudspeaker: "Truce period or no truce period, if you don't get the hell out of here, I'm going to shoot!" The two Egyptian vessels departed for Gaza and the Israeli ships followed closely. An hour later, Egyptian shore batteries in Gaza opened fire at the Israeli vessels, as Shulman had hoped. He radioed for permission to attack the Egyptian vessels, which were now at anchor. "No," came the response. Shulman radioed a second time, asking that his request be forwarded directly to Ben Gurion, who replied, "Paul, if you can sink them, shoot; if you can't, don't."[60]

Other sources don't mention a verbal warning being passed. In any case, it was dark by then. Positioning his squadron between the Egyptians and the moon, Shulman, as commander, put the four MTMs and Shayetet-13 in the water. Engines muffled, the crash boats headed for the enemy. Accelerating as they neared, and arming their explosives a hundred yards out, the pilots jumped free at the last minute.

A huge explosion; *Emir Farouk* had been hit. Not much later, a second assault boat scored a second hit on it. In flames, the enemy flagship sank within minutes. The third MTM steered into the minesweeper, and sank it too. The fourth, designated as a retrieval boat, pulled all three Shayetet pilots safely from the sea.[61]

"The sinking of the Farouk was Israel's most dramatic naval victory in the War of Independence. Some five hundred Egyptian sailors perished, many from that nation's upper class. However, the event received little formal publicity at the time: Israel wanted to draw no attention to its arguable violation of the truce; the Egyptians hoped to keep the Israeli triumph a secret. Nonetheless, news of the enormous loss reached the Egyptian public and for nearly a year the Egyptian navy had difficulty recruiting new sailors."[62] Another source states, "As the action took place twenty-four hours after a cease-fire, both sides kept the news a secret. The Egyptians were embarrassed to admit that their flagship had been sunk. The Israelis kept their sailors incommunicado, sailing up and down the coast between Haifa and Tel Aviv for five days."[63] Shulman had achieved what every commander seeks—tactical surprise—and dealt a heavy blow to a dangerous enemy, though breaking a truce in the process.

After this battle he was officially confirmed as Commander in Chief of the Navy, at age twenty-six.[64] He also commanded during a blockade of the Gaza Strip and the capture of Ein Gedi, securing the Dead Sea coast.[65] But though he had Ben-Gurion's support, he faced a political antagonist in Israel Galili of the Central Command, and indeed, much of the senior Haganah command structure.[66]

Although hundreds of Americans and other foreigners had crewed the surplus ships, friction sometimes developed between the Machal volunteers and those who had grown up in British Palestine. Their differences were exacerbated when, as in Shulman's case, the "foreigners" were seamen and the "natives" weren't.[67] Many land force leaders had served in the Jewish Brigade during the war, which contributed to what Shulman called a "British Army syndrome."[68] And finally, the army command tended to resent the resources Shulman asked for to defend the coast.[69] This friction was contained to some extent during the 1948 War, but re-emerged after victory.

In particular, the high command saw little need in the new armed forces for an American who didn't even speak Hebrew. When hostilities ended, even Ben-Gurion had to bow to political realities. He replaced Shulman as Commander with their mutual friend Shlomo Rabinovitch, fleeting the American up to be his own naval aide. Shulman subsequently served in various billets overseeing training, naval organization, tactics, and naval plans. But it became obvious that there was no real place in the new navy for one of its most influential founders,[70] and Shulman had his pride; seeing how the winds were blowing, he decided to step aside in March 1951.[71]

Meanwhile, ". . . .Egypt complained to the U.S. State Department that an American citizen had sunk its navy's flagship, and the State Department asked Shulman to resign his naval reserve commission."[72] Invited to explain, Shulman submitted documents attesting that he had never taken a formal oath or voted in an Israeli election, though admitting he had served in a foreign navy. Meanwhile, he had formed a civilian construction company. Eventually his U.S. passport was quietly renewed, and no further action seems to have been taken.[73]

Though he remained a U.S. citizen for the rest of his life, and made frequent visits Stateside, Shulman lived primarily in Israel for the next forty years. His companies drained swamps and built shipyards, airports, sewage

Shulman with David Ben-Gurion

disposal plants, listening posts, housing, roads, and the Tel Aviv City Hall.[74] He also served on the Board of Governors of the Technion, Israel's top university for science and engineering.[75] He lived a long, full life, and enjoyed sailing his 36-foot sailboat with his children and grandchildren.

The two Pan ships Shulman had bought for the "F--- Britain Shipping Company" went on to become the nucleus of Israel's merchant fleet.[76]

Shulman's daughter and grandson served in the Israeli Navy.[77]

Shulman died in 1994 in Haifa, from heart disease.[78] In 1995 he was posthumously promoted to "aluf" (admiral) by Prime Minister Yitzak Rabin.[79] At the promotion ceremony, the then-current Commander said that Shulman's ". . .own modesty cannot hide the fact that it was under his command our great navy was brought into being." [80] His picture hangs at the Clandestine Immigration

Paul Shulman '45's exhibit in the Levy Center at the U.S. Naval Academy

and Israeli Naval Museum in Haifa, along with those of the later commanders of the Israeli Navy.[81]

At USNA, there's an exhibit on him in the Levy Center.[82] He was also remembered in 1995 at a memorial service in Dahlgren Hall, where his widow Rose said, ". . . .his pride in having attended the Naval Academy never ended . . . He was part of the pages of the history of Israel."[83]

12

THOMAS J. HUDNER, JR, CLASS OF 1947: HE COULD BE COUNTED ON

The voice of USNA's only living Medal of Honor recipient is still strong, his recall still perfect at age eighty-seven. Silver-haired and straightbacked, Tom Hudner has a hundred stories from his four-decade flying career. To his credit, the ones he'd rather focus on are those that show him in a less than heroic light. (Such as the time he took off with his F-8E Crusader's wings still folded.)[1] What comes across loud and clear is a very old-fashioned word these days—*character*, from a time when schools and family and even boys' pulp entertainment focused on building men who could meet any test with quiet competence, selflessness, and honor.

Hudner's Irish great-grandfather was born in County Cork, emigrated in 1847, and settled in Fall River, Massachusetts. His ancestors on his mother's side came from Scotland late in the 1700s and settled in Lancaster County, Pennsylvania. Beginning as a street peddler, the Irish great-grandfather built his cart into a successful chain of meat and grocery stores, Hudner's Markets.[2] And aimed his family, like the Kennedys, at the upper reaches of society by sending his sons to prestigious prep schools and Ivy League universities.

Tom, born in 1924, was the oldest of five: boy, boy, boy, girl, boy. He remembers a "carefree existence" growing up, the stuff of New England memories. Summer meant a rental cottage at Westport Harbor, and sailing small boats on a freshwater pond. "It gave me a feeling for what sailing was. So (later) I was one of the few guys at USNA who actually had been sailing."[3]

Another experience that led him toward the sea took place in Rhode Island. "Since we lived close to Newport, we went there and saw the ships. They almost always had a destroyer or a sub there. I recall one sub there whose CO was the uncle of a close friend. Later he became COMSUBLANT and had quite a reputation. That visit would have been around 1938."

Another constant in Hudner's memories about growing up was his reading. Interestingly, the things he remembers is not great literature, but what used to be called boys' fiction. "I used to read about Captain Hornblower. I read all C. S. Forester's books . . . Captain Hornblower was a man of steel, yet very humble. I used to read G-8 and his Battle Aces too." This is an interesting choice of reading matter. "G-8" was a pulp-fiction flier and spy who battled German aces, vampires, werewolves, and other fantasy creatures in the sky over the trenches of WWI, and who now and then, if this author's memories of the series are still trustworthy, had to land between those trenches to pick up a shot-down and wounded comrade.

But the young Hudner's first love was athletics. "A brother of one of the guys in my group was a college football guy. That had some stature to our sandlot football team, the "Narrys"—for Narragansetts. And football opened some doors for us when we went to school."

Hudner went through those doors at Phillips Academy, a prestigious prep school in Andover, Massachusetts which his father and uncle had attended, and that two of his brothers would attend as well. He co-captained the track

Another Hudner Win
He doesn't always look this way

Hudner in the Philips
Academy yearbook

team, long legs propelling him through a hundred-yard sprint in 10.1 seconds. He played right halfback on the same team as Dick Boudin, played lacrosse, and was a senior class officer and student council member.[4]

Hudner was at Andover on December 7, 1941, when he heard the news of Pearl Harbor. "A friend in the class ahead of me at Andover enlisted in the V-5 program and got to fly PBMs, but got shot down. You might have heard of him—George Herbert Bush. He didn't inspire me to aviation, but I knew him . . . I just had a hankering for surface ships at the time. Airplanes just didn't interest me." Instead of aviation or Harvard, the young Hudner wanted to serve aboard a destroyer.

Hudner jokes, "My last game at Andover, it was a cold day. I got kicked in the cheek; it broke my cheekbone. They had a girls' school near us, and it crushed them all because I lost my looks."

He got his chance at destroyers on his Andover graduation in 1943, when Joe Martin, Speaker of the House of Representatives, appointed him as his second alternate to the U.S. Naval Academy. Tom was told to report to Annapolis on July 7, 1943.[5]

Hudner loved the Naval Academy. "I liked the discipline. And it was a different kind of person than used to be at

Andover. Andover was an elite school; USNA had some tough kids, not from the exclusive schools. Former sailors. A lot of them, had the circumstances been different, they could have been at Andover."

Due to the wartime acceleration, he spent only three years at the Academy. He played Jayvee football his third class and first class years and also lacrosse and track, though he didn't letter. His grades didn't exactly stand out. "I was 3/4 of the way down in my class. I was on the tree sometimes, when they posted our grades. 'On the tree'—that was what you called a 2.4, when the passing grade was 2.5." His best grades were in French; his worst, in "juice," or electrical engineering. "I did okay navigating. I enjoyed that."

Hudner's most vivid memory of his time at Bancroft was when he witnessed one of his classmates and teammates, "He had played football and had made the cut for the training tables," attempt suicide. "I was on a room in the third deck just above sick bay. This fellow jumped out of a tower window and landed on T-court. There was a letter in his room to his girlfriend. He had suffered a concussion the previous fall. He survived, however."

Hudner recalls the PCs (Patrol Coastal, subchasers) on which the midshipmen patrolled the Bay during the war. "We'd have a small crew and go out with a reserve ensign or jaygee as our CO. We had one jaygee who was very uncomfortable with these wise guys from the Academy. Everything we did he criticized. One day I was in charge of taking in the colors and he was in a foul mood. I dropped the colors overboard and then had to go and tell him . . . He told me, 'Go get it.' So I jumped overboard. That flag was so wet and heavy I almost didn't make it back aboard."

Hudner says, "I wish I could describe what happened when Japan surrendered. The fireworks. We were sitting on a cruiser a mile from Bancroft Hall. There was great excitement on the cruiser too, but I'm sure it was much more exciting ashore."

USS Helena

With the end of the war and '47's early graduation, the newly minted ensign put his chit in for destroyers. "All the real tigers in the class wanted that. It was the real Navy. But I was happy when I got a cruiser." He drew the newly commissioned USS *Helena*, CA-75, and joined up with her along with eight other new ensigns in Tsingtao, China. "The XO asked us what we'd like to do and most of us wanted deck division. He seemed to be pleading for a signal officer and I though, what the hell.

"It was the best job I ever had! A bunch of tough, gung-ho signalmen. These kids had fought a war to save civilization. I still get a card from one of them every year. Chief Douglas took me under his wing. Our first CO was McCollum (A. H. McCollum, USNA '21). He was in Naval Intelligence at Pearl Harbor leading up to World War Two. He was the one who deceived the Japs with the idea of the water message at Midway. He had us down to his mess two at a time; he was a real gentleman."

The new signal officer divided his time between the signal bridge and the radio spaces. "The crypto room was very small. My assignment was to type letter groups into the ECM, Electric Cipher Machine, to turn gobbledegook into clear language. We knew more than the rest of the people on the ship!"

Hudner was comfortable aboard the cruiser, but when they got back to Long Beach, orders were waiting for him to go to Pearl as a communications officer. "All the reserve communicators were getting out so I went to the CINCPAC/CINCPOA staff. We worked hard but had a lot of fun. There were four of us ensigns. All the captains' daughters were about our age and all of them were pretty good looking."

As Hudner tells it, going Navy Air was not his idea. "These other ensigns were off carriers and wanted to go Navy Air, but I wasn't interested. I wanted to be at sea." Hudner had worked out a swap with an ensign aboard DD-888, USS *Stickell*, with that skipper's okay. But at the same time, the CNO was soliciting applications for flight school. "The other ensigns shamed me into putting my chit in." As it turned out, the man he was relieving transferred off *Stickell* just before Hudner's orders to flight school arrived, leaving the destroyer to put to sea short an officer. "The captain, Jack Chew (John Lewis Chew, '31, buried at the USNA Cemetery)[6], was furious. The next time I ever saw him was at a black tie affair at the Shoreham Hotel, when he was a three star admiral. 'I don't think you remember me,' I said. 'Oh, yes, I do,' he said, glaring at me."

The putative pilot arrived at Pensacola in mid-April 1948 for flight training. "We did our basic in SNJs, then went to Corpus Christi for Advanced in F4Us—Corsairs. The Corsair was a good airplane. By some comparisons, better than the F-51. The sexy airplane of the Fleet—a lot of power, but the nose was so long you can't see on carrier approach. You had to fly in sideways to see the LSO." Hudner received his wings of gold on August 12, 1949, at Cabaniss Field.[7] He was first assigned to a Skyraider squadron, then reassigned to Fighter Squadron 32, which had F8Fs. "But when I got there they were transitioning to Corsairs."

Hudner joined Fighter Squadron 32 in November of 1949.⁸ The next May, the squadron deployed from Quonset Point aboard USS *Leyte* (CV-32) as part of Air Group Three for a routine Med deployment. *Leyte* was anchored off Naples when word came that the North Koreans had invaded South Korea, but orders to the Pacific didn't arrive until August 8, when she was in Beirut. The carrier and embarked air group returned to the States en route to the Pacific via the Panama Canal, and arrived off Korea for operations on 10 October, three weeks after the landing at Inchon.⁹ From then through 19 January 1951, *Leyte* and Air Group Three would spend ninety-two days at sea and fly 3,933 sorties against the North Koreans.¹⁰ One of those sorties, on 4 December, would make Hudner famous . . . though at a cost he would rather not have paid.

In the mountains around the Chosin reservoir, 15,000 Marines were surrounded by somewhere near ten times

Group of Corsairs

USS Leyte *anchored at Sasebo, 1950*

USS Leyte *operates with a ZPG-2 airship, during ASWEX 1-58, Feb 1958*

A PC-461 class sub chaser

their number of Chinese forces, who had surreptitiously filtered into North Korea as US and ROK forces neared the Chinese border. They were trying to break out though a narrow mountain road covered by up to two feet of snow, with nighttime temperatures down to minus thirty-five degrees Fahrenheit. It was a desperate situation, and the airmen of Task Force 77, built around USS *Philippine Sea* (CV-47), USS *Missouri* (BB-63) and USS *Leyte*, had been flying with little rest for several days to lend close air support.[11]

At 1338 that afternoon VF-32's exec, Lcdr Dick Cavolli led an armed recon mission off *Leyte*'s deck.[12] The mission included Cavolli's wingman, George Hudson, section leader Ensign Jesse Brown, the Navy's first African American combat aviator,[13] and Lt (jg) Tom Hudner.[14] Two more pilots brought up the rear, making a flight of six Corsairs in all. Hudner knew all his fellow pilots, of course. "I first met Jesse in the locker room when I was changing for a flight. He was a friendly person, someone who right away was the type of person you knew you would like."[15]

Ensign Jesse Leroy Brown

Probably unknown to Hudner then, Jesse Leroy Brown had already proven his own heroism many times over, overcoming enormous obstacles simply to be able to change in that locker room with other pilots. Born of sharecroppers in Hattiesburg, Mississippi, in 1926—a town that had seen nine lynchings since 1890—Brown had caught the flying bug watching cropdusters and barnstormers fly from the local dirt strip. Howard Hughes's 1930 film *Hell's Angels* had sealed his ambition—a nearly insane one for a black child in that era's Mississippi—to fly. Brown had gone north as soon as he graduated from high school, and worked his way through Ohio State as a janitor and loading boxcars at night. The moustached, boyish-looking Brown had endured slurs and racism in Pensacola, winning his wings in a Navy that up to then had assigned black men as cooks, messmen, or stewards—not as officers and pilots in combat.[16] By now Brown had flown nineteen combat missions and had already been awarded the Air Medal for pressing home attacks despite hostile antiaircraft fire.[17]

After the takeoff in the gray, overcast afternoon, heavy with 500-pound bombs, HVAR rockets, a 150-gallon external fuel tank, napalm pods, and 2400 rounds each of .50-cal,[18] the flight headed for the mountains. The lightly-armed Chinese had little in the way of anti-aircraft artillery, but on the sixty-nine previous sorties to the reservoir the pilots had encountered heavy and coordinated small-arms

JLB playing Acey-Deucey in Ready Room aboard USS Leyte, 1949

fire that had left bullet holes in several aircraft. Earlier that day rifle fire had brought down a Marine Corsair; the pilot had crash-landed and been killed. VF-32's skipper had warned his pilots not to go in too low.[19]

The Corsairs covered the fifty miles from *Leyte* to shore. The recce area around the reservoir was another fifty miles inland, over mountains rising to six thousand five hundred feet. The planes flew at a thousand feet above that, and the cold quickly seeped into the cockpit and through their long johns, poopy suits, fleece-lined boots and plastic football helmets with holes cut for headphones.[20] At a little after 1430, the flight descended to five hundred feet, low and slow, along the west side of the Chosin to search for targets on the ground. Minutes later, one of the tail-end pilots noticed a stream of vapor coming from Brown's Corsair. He radioed him, "You're dumping fuel."[21]

Ensign Brown in the cockpit of F4U-4 Corsair

Hudner says, "... the first I knew of Jesse's trouble was when he came over the radio."

Brown reported he was hit and losing oil pressure, which would stop his engine very quickly. "No Mayday, no panic. Just a calm announcement from Jesse that he was going down."[22] Cavolli told Hudner to look around for a crash landing site, but Brown had already seen one, a bare patch on a mountainside, and was headed for it. "I went over with him a checklist of the things he should do, such as opening the canopy ... When he did land, it was without power so he did not have much control. He hit with such force that the fuselage buckled at the cockpit and his canopy was slammed shut because of the force of the landing. We didn't think there was any possibility of survival ... After circling several times, we noticed that he had opened the canopy and was waving at us to let us know he was alive, but he did not get out of the cockpit."[23]

Cavolli climbed to improve his radio transmission and called back to *Leyte* for a rescue helo. Meanwhile, the orbiting pilots observed smoke seeping from the cowling of the crashed Corsair. "We couldn't understand why Jesse didn't get out of his Corsair," Hudner wrote. The radio call came back that the helo would be there in twenty minutes. Orbiting, he became convinced that Brown's plane was about to burst into flames, with certain death for a pilot somehow trapped in his cockpit.[24]

"Knowing that help was on the way but that time was becoming critical, I decided to make a wheels-up landing, pull him out of the cockpit and wait for the helicopter . . . I made this decision without informing Cevoli, who was on another channel. I don't know whether he'd have given me permission to land, but I doubt it. I let down, fired off my rockets and ammunition into the hillside and slowed to 85 knots indicated airspeed . . . I made a pass over Jesse's plane with my wheels up and flaps down to get the feel of a carrier-type approach, then circled and let down close to the ground with the idea of flying up onto the slope to minimize the impact.

"I hit hard. The ground under the foot of snow was like cement . . . the windshield cracked, probably because it was so brittle from the cold and the force of the impact."[25] Hudner had also injured his back in the crash landing, which he later called "The hardest landing I have ever had."[26] He exited his aircraft and trudged the hundred yards to Brown's plane. When he got there, he saw why his wingman was still in the cockpit. "Because of the way the plane buckled, it caught his leg in between the side of the fuselage and a hydraulic control panel under the instrument panel. It pinned him in so that he could not move."[27]

The following hour was both tense and increasingly frustrating. As the other four planes in the flight orbited, ready to drop hell on any enemies who approached the

Painting of JLB's crash landing by Matt Hall

double crash site, Hudner went back to his own plane and called the helicopter to bring along a fire extinguisher and an axe. Meanwhile, he threw snow on the fire and put his watch cap and scarf on the trapped pilot.

When the helo arrived, Hudner recognized 1st Lt. Charlie Ward, one of the Marine pilots Leyte had ferried to from Norfolk to Japan.[28] Ward's helo had bad brakes, not a plus on a steep mountainside, but at last he got it down.[29] When Hudner and Ward got to Brown, they found they couldn't get to the fire to extinguish it, and the axe just bounced off the fuselage. They tried to climb up the wing and fuselage to where they could lever Brown out of the crumpled cockpit—as it was, he was high above their heads—but slipped and fell each time they tried. Meanwhile, Brown was drifting in and out of consciousness, both from the freezing temperatures and blood loss from his leg, which the others couldn't get to. "There was absolutely no panic in his voice. His attitude was one of resignation. His manner gave me inspiration."[30]

Meanwhile, the early winter dusk was drawing on. Hudner says, "The helo pilot and I, in our emotion and panic, and with the light of day fading, discussed using a knife to cut off Jesse's entrapped leg. Neither of us really could have done it, and it was obvious Jesse was dying. He was beyond help at that point."[31]

Finally Ward, whose helo was not equipped to fly at night, said they had to leave. With great reluctance, Hudner realized his only choices were to fly out with Ward, or die uselessly along with Brown, who was no longer responding. "Darkness was setting in and we'd never get out after dark. We had no choice but to leave him. I was devastated emotionally. In those seconds of our indecision, Jesse died."[32]

It took Hudner several eventful days to make his way back to *Leyte* through bad weather and the chaos of the UN retreat.[33] Meanwhile, a Chinese fleet was reported heading toward Formosa, and an intelligence report indicated that the Soviets were preparing an all-out attack on Japan.[34] On 6 December, in view of possible contingencies, the Joint Chiefs of Staff sent out a general alarm to American forces throughout the world. When he did arrive back aboard, Hudner advised the carrier's CO to incinerate Brown's aircraft and body with napalm rather than risking more lives to return his body from what was now Chinese-held territory. This was done on December 7.

There was no discussion on the ship about anyone disciplining Hudner. "Jesse was too well loved for anyone in the ship to have entertained the idea. I do know, though, that there were those not on the ship who did think I should have been severely punished."[35]

Instead, however, he was awarded the nation's highest award for his selfless act. He received the Medal of Honor from President Harry S. Truman at the White

House on April 13, 1951, the first presentation of the Medal since the end of WWII. The armed services had been integrated only since the previous year;[36] though Hudner's courage certainly deserved such an award, it is possible that Truman's approval of the Medal of Honor for a white man who risked his life rescuing a black man might also have reflected the President's personal investment in the success of military desegregation. Hudner himself says: "Whether the decision to award me the Medal (was) because of the color factor, I don't know. I'd like to think it didn't. But I'm sure there were those who wanted to take advantage of the situation, especially since the war was not going too well for us at the time."[37] Present were the CNO, ADM Forrest Sherman, Secretary of Defense George Marshall, and Mrs. Daisy Brown, Jesse's wife.[38]

Brown was posthumously awarded the Distinguished Flying Cross and Purple Heart, and USS *Jesse L Brown*, DE-1089, was christened in 1972. Charles Ward, the helo pilot, received the Silver Star for his part in the action.

Captain Tom Hudner retired from active duty in 1973. After some years in private industry, he became the deputy commissioner of Veterans' Services for the Commonwealth of Massachusetts. In 1991 Governor Dukakis appointed him Commissioner of Veterans' Services. In 2006 Hudner received the Alumni Association's Distinguished Graduate Award in a ceremony before the Brigade in Alumni Hall.[39]

After fully retiring, Hudner lived in Concord, Massachusetts with his wife Georgea, whom he met in San Diego after her previous husband, also a naval aviator, had been killed in a traffic accident. They shared the bringing-up of Georgea's three children by her first marriage, and had one more child together. "We are a close family," Hudner said.

What would he say to today's mids? The question doesn't slow him down for a millisecond. "I'd just say, to today's plebes: Study hard, but remember the responsibility you're

Hudner being awarded the Medal of Honor in April 1951

Hudner with President Harry S. Truman after receiving Medal of Honor

Hudner wearing Medal of Honor

Captain Thomas J. Hudner in 1972 at launching of the Jesse L Brown

Hudner in 2010, courtesy of National Naval Aviation

taking on to become a naval officer. Integrity. Responsibility. Separate the wheat from the chaff that you hear on radio and tv and remember where we came from as Americans, and what we inherited from our country's founders."

Thomas J. Hudner passed in 2017, shortly after helping with this book.

13

JOHN RIPLEY, CLASS OF 1962: "HOLD AND DIE"

"If a young officer or Marine ever asks, what is the meaning of Semper Fidelis, tell them my story." —John Ripley

There's no other silence at the Academy like the hush that hovers over the parquet floor in Memorial Hall. Plaques and memorials hang beneath murals of engagements of our wars. Here we remember those who exemplified the highest traditions we aspire to. Who—literally—gave their lives for our country.

But even of them, heroes all, how vanishingly few ever received the order every military professional dreads. To hold your post in a hopeless situation and buy time for others with the wager of your own life . . . to hold and die.

Most grads will remember, in a niche in the entrance, a glassed-in diorama depicting a man swinging by his hands beneath a trestle bridge while explosions fill the air. That display, commissioned by the Class of 1962, was executed by Robert H. Mouat, a retired Royal Navy Commander.[1] The bridge was at Dong Ha, in Quang Tri Province, South Vietnam, where in 1972, one man stopped two enemy divisions, and saved a country that could not save itself.

That man was John Ripley, Class of '62. His story's one of resolute faith, personal courage, leadership, and achievement so superhuman it's become a legend.

There's a temptation, when confronted by such a hero, to take one of two opposite tacks. The first is to portray him as flawless, superhuman, almost a demigod. The second is to suspect such a paragon could not exist, and relegate his story to exaggeration or myth.

We must steer between Scylla and Charybdis to find the truth.

John Walter Ripley was born in 1939, and grew up in Radford, Virginia. His father, Francis Droit Ripley, was a gruff, cigar-smoking engineer, who eventually became a manager for the Norfolk and Western Railroad, of which his own father had been a vice-president.[2]

But there's a backstory. Francis "Bud" Ripley had been appointed to USNA with the Class of 1922, but resigned at the beginning of his first-class year. His Academy records document a recurrent pattern of conduct deficiencies. We won't detail them, but the last straw seems to have been when Ripley went on unauthorized absence for five days from Kansas in Seattle, after an earlier UA episode in Honolulu. "The SecNav accepted his resignation "for the good of the Service" on August 24, effective September 1, 1920."[3]

Perhaps to compensate for what he'd recognized as missing the boat in more ways than one, Bud Ripley always emphasized his family's martial heritage.[4] His ancestors had fought in most of America's wars; his wife Verna's grandfather had served in the Virginia Light Artillery and seems to have fought at Gettysburg, in Rodes's Division.[5]

Bud was also an intensely religious Catholic, in a strict mode. The family said the rosary each day, kneeling before a makeshift shrine centered on a crucifix hung with portraits of Robert E. Lee and Stonewall Jackson on either side. Even after Vatican II had changed the Mass's language to English, Bud still gave his responses in church in Latin, and at an uncomfortably loud level.[6]

Francis's grandson Stephen says, "He (Francis) had a rocky career and at points my father had to live with his grandparents ... Dad's upbringing was tough. His family while very religious was not the most cohesive ... Dad grew up literally on the wrong side of the tracks. His house was between the railroad and the New River. He partially raised himself and lived outdoors on the river. He was as wild as they come. I think this developed in him a powerful sense of self-reliance and toughness."[7]

Seventeen-year-old John joined the Marines out of high school, inspired by selling newspapers on troop trains to Marines and by reading Leon Uris's novel, *Battle Cry*. After he joined, Francis pushed him hard to attend the Naval Academy. He locked John in his room to force him to study for the exams.[8] A year after enlisting, he was approved to attend the Naval Academy Prep School (NAPS).[9] He was sworn in at the age of 19, on June 30, 1958.[10]

The Academy of 1958-62 may have shared some of the buildings of the current Academy, but many other things were quite different from the institution mids know today. As a former enlisted Marine, Ripley seemed very "squared away" to his company mates.[11] A tougher struggle for him were the rigorous academics, heavy on science, math, and engineering.[12] According to the archives, Ripley qualified as Expert Rifleman and Expert Pistol on August 29, 1958. On September 29, 1961, he received seventy-five demerits and was restricted to quarters for thirty days for being "Absent over liberty for a period of fifty minutes and failing to notify the Officer of the Watch on his return." He graduated 779th out of 789.[13]

Ripley's plebe summer friend, and one of his roommates second and first class year, Pete Golwas, remembers a lot more detail than those bare-bones records. "John had a lot of what I call texture. He was a very complicated, focused guy, a whole guy," Golwas says. "I first met John during plebe summer. One of my roommates was one of his friends and he hung around our room a lot. From the beginning, his stated ambition was to be the Commandant of the Marine Corps. So we (John and I) sort of made it our mission to make it impossible for him to become the Commandant!

"First class year we were roommates in one of the few five-man rooms on the top deck of the sixth wing (facing Luce Hall). It was me and Ripley, and Minor Carter, Steve Todd, and Tony Zaccagnino; we were all athletes, track, baseball, boxing. John never did any varsity sports, but he was a good athlete; he played battalion football and did some other extracurricular stuff.

"By first class year the two of us were maxed out on demerits. But still, we went over the wall to chase girls at the University of Maryland, College Park. We went over the wall a lot—I would say three to four times during our first class year, when we were on restriction. We went around to the water side of Halsey Hall and snuck out around the chain link fence there. That fence had a couple of little links that kept it taut; we would shimmy around the end of it, hanging on out over the water, and get to the other side. Generally, we'd get in contact with some girls who had cars and would take us where we wanted to go. Once we went up the night before a big football game at Maryland, and we would go to the frat parties there where we could drink for free."

One time they got caught coming back. "For some reason, we thought we could get back in through those big bronze doors in front of Memorial Hall. And about 11 PM the OOD heard us pounding on the doors. (We'd had a few beers.) He came out and fried us on the spot. We were already maxed out, but we didn't have the sense to realize the possible repercussions. But I was captain of the track team, and I think one of our company officers took a shine to us and made something happen for us.

"John and I both graduated on probation because we didn't have any damn sense. But we graduated . . . John was very close to being the anchor man. He had a tough time with science and math. Though his essays were all first class; he got a solid A+ on his First Class essay about the occupation of the banana republics after the Spanish-American war. Calc and Skinny and Steam were his weak points. But he burned the midnight oil; he knew he was up against it and he worked hard. The stuff that had to do with science, he just couldn't do. But his roommate Tony Zaccagnino was a star man (and undefeated boxer)—a little guy, but a stud, he could kick the shit out of people—and John would be studying, and if he came to a place he couldn't understand, he'd ask Tony for help.

"John was one of the most focused people I've ever known, but one thing that endeared him to me was that he was a risk taker, and we'd take risks together . . . One time we were on liberty in Philadelphia together, after taking the damage control course there, second class summer. We were broke and wandering around in our trop white long looking for free beer. So we went to the worst part of the city, Skid Row, and went to a place that bought blood. We called that "vampire liberty" in those days. His blood was RH negative, so he got seven dollars a pint. Mine was O positive so I only got five dollars. He got more money than I did, so he got to drink more than I did!

"During the summer between second class and first class year, John and I took liberty in Washington, DC. We wanted to drive from Washington to my home in Arkansas and back along the Gulf Coast and Florida. So we bought a junker car—paid $150 for it, and it ran maybe a mile before it quit. It was a wreck. Mr. Carter (Minor Carter's father) loaned us his car and it ran. We stopped at Apalachicola Florida on the way back and bought raw oysters. We each thought the other guy had the money. But when it came time to pay, neither of us had any. So we had to wash dishes for two days before they let us go. The guy let us stay free overnight, so he could work us some more the next day!

"Another time, the night before graduation, we took our girls—his girl was Moline, who would become his wife—they both wanted to see a mid's room. We didn't mention that it was against the law, but they must have suspected. We put both girls in a "moke cart" (*Au*—mid slang for a laundry cart) and put dirty laundry on top of them and took them up to the sixth floor in the freight elevator. They thought the room was pretty nice.

"John was a good Catholic; I would even call him a devout Catholic, but he was never seminary material. We generally sat together at Mass. He thought his faith would help him to be Commandant of the Marine Corps; there were a lot of Irish Catholic commandants."[14]

Another of John's classmates in that famous room, Minor Carter, says, "To this day our room is renowned for having the best time at the Academy. I first met John plebe year. He was very squared away, had a good sense of humor. He, Zack, Pete, and I lived together most of four years.

"During third class year, U of Maryland came down and stole the goat. Next day Admin said, 'You can go out into the Yard and guard the Yard.' I guess they thought Maryland would try something else. Our room decided the best defense was a good offense. So we went over the wall back of Halsey and a girl picked us up and took us—Pete, Zack, Rip, Tony Z and me—to College Park.

"Now, on fraternity row there, they had a road that wound around a patch of grass to make a bowl. We poured gasoline on that grass in the shape of an N, and lit it. It made a nice big black N. It didn't seem to make them mad. We partied a little with the fraternities there and headed back. We sneak back over the wall, at the same place, and head to the third wing. As we come up the ladder and push the door open, there's Inside Formation, and they're taking muster. John looks at Pete and says, 'This could be trouble.'

Pete says, 'Really?' And the rest of us are going, 'Oh, Shit. This is a problem.' We each got twenty hours of extra duty. In those days you got up in sweat gear at 0515 and walked the curves and jogged the straights at the Field House, and came back at 0615 to get ready for reveille.

"During second class year we were in this five-man room in Sunshine Alley. That's the passageway in fourth wing that fronts on T Court; there's a square in the middle and the P-way on T Court only has rooms on one side; the other side is all windows. It's about 1800. We're all there and I'm about to take a shower and I hear this scuffle outside. Then I hear Pete and Ripley yell 'Give us a hand.' So I run out and there's this plebe they couldn't stand for reasons that are unclear even today, and Ripley said, 'This guy has to go in the shower.' So we had to throw him in the shower. He fought, though—it took three of us. The plebe was on his way to watch squad inspection; we never found out what happened there.

"This was at the end of the Dark Ages. And there was a room full of firsties right next to us then; they were not just in another company, they were in another battalion. But on the morning of Graduation Day one day at reveille we hear these wild screams and cursing. We came out in the P-way, wondering what in hell was going on. Well, a bunch of plebes had filled those big trash cans full of water and had gone in and destroyed the firsties' room. Naturally the firsties were going nuts and the plebes responding in kind. And when we went out, the plebe who we'd thrown in the shower sees us and goes pale, and says, 'Oh, shit, guys . . . we got the wrong room.'

Another story from Carter: "At the end of second class year this firstie walks in and says, 'Are you John Ripley?' And John says 'Yes.' 'Are you a size 42?' 'Yes,' says John, a little puzzled.

"If you don't pass your Wires re-exam, can I buy your trop khakis?"

We were all laughing at him. 'John, they must have a lot of faith in you,' somebody said. 'They want to buy

Ripley in the 1962 Lucky Bag

your uniforms.' But he passed that re-exam, fortunately."[15]

Finally, Ray Madonna, 62's former class president, added his take. "My first impression was that he seemed like a normal mid. In good shape, average height, dark short hair—he carried himself well. We both became part of a kind of eclectic group from across the Brigade whose common interests were, I guess, mainly sports and parties.

"He had his share of issues with the system, as most of us did who ran around in this group. One of our company officers told me once, nobody gets away with things forever.

"One time during second class year he and I were at a certain party, and there might have been some illegal use of alcohol. John and I got called in and asked if we'd attended a certain party we shouldn't have been at. I don't know what he said—we were talked to separately—but he got just as many demerits as I did. Honor was important to him; I'm sure he came right out and 'fessed up. We both did our share of "walking tours" with the M1 and in later years, mustering in every hour at the Main Office. I graduated on conduct probation and I think he did too.

"John did a 180-degree change right after graduation. We all got thirty days' basket leave after graduation. But Ripley gave that up and reported in to Basic before the rest of us . . . I don't think he ever broke a reg after that. Maybe it was a Marine thing!"[16]

Ripley with Marines in Vietnam

Ripley's first active duty assignment was afloat, with the MarDet aboard *Independence* (CV-62). After this he joined 2nd Battalion, 2nd Marines, first commanding a rifle, then a weapons platoon. In May 1965 he went to 2nd Force Reconnaissance Company, where he completed Airborne, Scuba, Ranger, and Jumpmaster courses.[17]

In 1965, Ripley deployed to the Dominican Republic when the U.S. decided to intervene in the civil war there.[18] Marines landed in Santo Domingo and assisted in the evacuation of 1300 civilians there.[19] There was some fighting, but only a taste, and the Marines were withdrawn after six weeks.

In the summer of '66, Ripley got his orders to deploy to Vietnam.

He joined 3rd Battalion, 3rd Marines, as commander of Lima Company.[20] The 3rd MarDiv had responsibility for both Quang Tri and Thua Thien provinces, operating from bases south of the DMZ, to deny the enemy access to the lowlands.[21] The NVA, on the other hand, wanted to keep U.S. troops away from their own logistic trails and supply routes. By 1966 this struggle was at its fiercest. Marines out of Camp Carroll, Khe Sanh Combat base, the Rockpile, Vandegrift Combat Base and in operations around Dong Ha, Cam Lo, the A Shau Valley, Con Thien, and Da Nang (the "Leatherneck Square") were in heavy action almost daily.[22]

Lima Company, "Ripley's Raiders," carried out patrolling, to clear villes, set up ambushes, find the enemy, fix him, and destroy him.[23] Their commander was in the thick of it. The official description of the action for which he was awarded the Silver Star reads:

> On 21 August 1967, Company L was assigned the mission of reinforcing a convoy that had been surprised by a large enemy force and was pinned down. With one rifle platoon, a small command group, and accompanied by two M-42 dual 40-mm. anti-aircraft guns, Captain Ripley was leading the relief column when it suddenly came under intense enemy automatic weapons and recoilless rifle fire. Disregarding his own safety and the heavy volume of hostile fire, he moved to the machine gun mounted on the vehicle and opened fire, pinpointing the location of the well concealed North Vietnamese and enabling the 40-mm. guns to deliver accurate fire on the enemy positions. Directing his unit to dismount, he quickly organized a defensive perimeter while coordinating supporting artillery fire and simultaneously controlling the remainder of his company which was widely separated from his position. Repeatedly exposing himself to the hostile fire, he directed artillery fire and air strikes upon the attacking enemy force and courageously adjusted fire missions to within fifty meters of his position. Throughout the following three hours, his skillful employment of supporting arms and direction of the fire of his men repulsed the determined enemy attacks and forced the hostile units to flee in panic and confusion...."[24]

Though wounded in one night-long action that left half the company killed or wounded,[25] Ripley returned to resume command in March 1967 ... at about the same time Navy Seabees were building a two-lane highway bridge across the Cu Viet River, in the little village of Dong Ha.[26]

His tour ended, he returned to the U.S. After attending Amphibious Warfare School Ripley became the infantry officers monitor, headquarters, Marine Corps. Then, selected as exchange officer to the British Royal Marines, he attended the Commando Course at Lympstone. He served in Singapore with the 3rd Commando Brigade and with 40 Commando in Northern Malaya, campaigning for several months with the famous Gurkha Rifles. He also trained with the Special Boat Service in northern Norway.[27] And then, he volunteered for his second tour in Vietnam.

The most thorough treatment of the defining moment of Ripley's life is *The Bridge at Dong Ha*, by John Grider Miller, who based it on personal interviews with Ripley and others involved in the action. The discussion below is based on this excellent and readable book from the Naval Institute Press.[28]

By 1972, the half-million troops the U.S. had committed in Vietnam had been largely withdrawn, leaving primarily supporting arms liaisons and co-vans, U.S. military advisors to South Vietnamese commanders. Hanoi was ready to finish the war. The Easter Offensive would punch two armored divisions across the DMZ, led by the elite NVA 308th, south down Route 1 to Saigon and victory. To get there, they would have to seize the highway bridge at Dong Ha. On Good Friday, 1972, the seven hundred men of Third Battalion, Vietnamese Marine Corps, detrucked at Dong Ha with orders to hold the bridge. Their commander was a much-decorated, much-wounded Major Le Ba Binh; and Binh's co-van was thirty-two-year-old Captain John Ripley.

As dawn approached, artillery began to rain down on the town. Ripley, who'd spent the night trying to sleep under shelling in the abandoned Camp Carroll base morgue, was getting ready to head to the bridge with Binh when the radio message came in, "Expect enemy tanks." Both men were taken aback. Binh's men had not been trained to stop armor. They had LAAWs—light bazooka-like weapons—but only ten of them, and each could be used only once.

As they neared Dong Ha, where Binh's Marines were digging in, thirty to forty thousand civilian refugees were streaming down the road, along with deserters from the Army of the Republic of Vietnam forces deployed farther north. Forty ARVN medium tanks also joined them. But then a spotter plane reported that over two hundred enemy tanks, heavy Russian T54s, where heading down Highway 1 toward the bridge. After a short discussion with Higher, Ripley decided the bridge could not be held with the forces available. It would have to be destroyed, and he asked for demo charges to be sent right away.

The M48 column proceeded along the river, stopping NVA infantry from crossing on an old railroad bridge en route, but when heavy shelling prevented them from going through Dong Ha Ripley and Binh took a back road. Cresting a hill, Ripley raised his binoculars. There was the enemy: green tanks, red stars, rising dust as they raced toward the bridge, but not yet head on. A moment later the lead ARVN tanks fired. The first three NVA tanks disintegrated from direct hits on their thinner side armor. But rumors and panic were spreading over the radio, and the southern tank commander was losing his courage.

Half an hour later Binh and Ripley looked down on the long steel bridge as a lone, very brave Vietnamese Marine sergeant, dragging dirt-filled ammo boxes as cover, crawled out onto it. As the first T-54 edged up onto the bridge, the VM fired two M-72 LAWS and jammed its turret. The tank backed off. The VMs covered Ripley and the tankers' co-van, U.S. Army major Jim Smock, as they made a desperate sprint under fire down the open hillside to the bridge abutment.

They found demo charges there, TNT in wooden boxes and C4 in satchels, but the ARVN engineers sent with them had disappeared. Ripley sized up the bridge looming above them. It was well built, a massive, steel structure on reinforced concrete piers. After Ranger school, he knew, at least theoretically, how to destroy it. He could see six steel I-beam stringers, the vertical part of the I about a yard high, supporting the roadbed. He explained the plan to Smock. The TNT should fit into the channel between the stringers, held in place by the lower flanges of the I beams.

Ripley said, "We can load boxes into the channels sideways and I can drag them out over the water. I'll start in the downstream channel and take the first box out to the first pier. Then I'll position the rest of the boxes in the other channels along a diagonal that runs back toward this shoreline. I can use the satchel charges—in those haversacks—to set up twisters on each girder. They'll cut the I-beams and detonate the TNT, which'll torque the near span right into the drink." He asked Smock to look for detonators, but first, to hold up the razor wire protecting the abutment from saboteurs so he could get through and start crawling.

Ripley got hold of the lower stringer and managed to pull himself over and through the razor wire, but at the cost of a lot of slashes on his legs. Hanging by his arms, he began to work his way out to the pier. Small arms fire still crackled over the river from both sides. Then the tanks began exchanging fire. But no more NVA tanks tried to cross, and no one seemed to have noticed him yet, hanging beneath the bridge like a spider in a doorway.

He reached the pier, twisted himself up between the stringers, and placed two of the satchel charges. Then, crawling within the narrow channel of the two I-beams, knees and fingers scraping in the rough flanges, he worked his way back to the shore. There Smock had loaded two boxes of TNT and two more satchels into the channel. Still cramp-locked between the I-beams, Ripley backed a hundred feet out to the concrete pier again, dragging a hundred eighty pounds of explosives, yanking them painfully when they snagged on rough metal. When they were set, he had to drop out from the first channel, brachiate to the next one, and muscle himself back up into the nestling steel. When he exposed himself the NVA began shooting, some from beneath the bridge, but none hit him. At least not that time.

Back to the shore. He made the cramped crawl, backward, dragging explosives, five more times, increasingly faint from loss of blood and sheer exhaustion. He pulled himself through the razor wire again and collapsed. But then, though the charges were in place, there were still no detonators.

Electric ones would have been ideal, but there didn't seem to be any. However, Smock had found blasting caps and primer cord. On the other hand, they had no pliers to crimp the caps to the cord.

Lying on the river bank, Ripley broke every safety rule in the book, field-measuring fuze by wrapping it around his chest, then crimping the caps to the cord with his teeth, a procedure that would have blown his head apart if he'd bitten down even a half-inch wrong.

This time when Smock pulled down the wire to let him crawl through, the NVA began firing heavily. Explosive caps and fuze buttoned into his pockets, Ripley again had to monkey-bar himself out a hundred feet, hanging by his arms, while every rifleman on the far bank tried to hit him. He chanted rhythmically to keep up his momentum. "Je-sus Mar-ry get me there. Jesus-Mary-get-me-there. JesusMaryget me there. . . ." He reached the first charge and had just twisted himself up into the protection of the channel when a tank shell hit the stringer less than two feet from his hand. It didn't go off, but the impact almost knocked him out, which would have meant a long fall into the river.

"Ripley at the Bridge" Diorama in Bancroft Hall

He clung, though, and went to work with his Ka-Bar, digging a pit for the cap in the creamy C4, then wedging a damp-resistant C-ration match into the end of the fuze before lighting it with another. Then swung to the next charge. When the fuzes were burning, sputtering a sparkling fire beneath the bridge, he swung himself back to the same cadenced out-loud prayer. Once again, amazingly, none of the hundreds of bullets fired at him found its target.

He got back this time, and collapsed in exhaustion and pain . . . to find Smock waving a new find: electric detonators. Ripley looked back at the burning fuzes. He'd cut them for thirty minutes. He didn't know how long was left. But fuzes could fail.

"I have to go back out on the bridge," he said. "Got to set the backup."

Smock tried to argue him out of it, but gave up when Ripley began pulling up comm wire and coiling it to carry. Once again, John Ripley crawled hand over hand out onto that bridge, all the way, and this time stayed out there, cutting wire, inserting detonators, taping splices, while the bullets cracked around him and below him the fuzes, growing steadily shorter, burned away. At one point he lost consciousness and nearly fell again, but recovered and worked on. And still the NVA didn't rush the bridge, and no more tanks rolled up to try to cross.

He finished. Hand over hand back to the shore. Still not hit. Through the razor wire. Smock, shouting and praying aloud, grabbed him and rushed him up the slope as Ripley paid out wire and the VMs chanted *"Dai-uy Dien."*—"Captain Crazy."

Uphill, Ripley got briefly sidetracked trying to help a lost Vietnamese child, then attempted to connect the wires to the battery of a burning Jeep. Try as he might, he couldn't get this field-expedient hookup to work. From across the river came the growl of diesels starting again.

With a shattering roar, the bridge exploded. The shock wave slammed him into the ground. Beams and debris flew hundreds of feet into the air, succeeded by a gray pall of smoke. The time fuzes had worked.

Lying propped against a bunker, being fed a can of condensed milk, Ripley watched A-1 Skyraiders and destroyer gunfire take their toll of the stalled, backed-up enemy columns. Against all odds, the Easter Offensive had failed, stopped by one brave and utterly determined man. John Ripley had entered the pantheon of heroes.

Ripley was awarded the nation's second highest honor, the Navy Cross, for his actions at Dong Ha. He returned from Vietnam to serve as the Marine Officer Instructor at Oregon State. In 1975 he attended American University for postgraduate work, after which he was assigned to the Office of the Chief of Staff, HQMC, as the Administrative Assistant/Aide to the Chief of Staff.

Ripley next commanded 1st Battalion, 2nd Marines. On completion of this tour, he attended the Naval War College. He served with the Joint Staff, Joint Chiefs of Staff, before being assigned to the U.S. Naval Academy. Grads from the 1984-87 time frame may remember him as the Director, Division of English and History; the present writer first met him there during this period. Ripley next spent a year as the Assistant Chief of Staff, G-3 with the 3rd Marine Expeditionary Force, Okinawa, Japan.

That was John Ripley, the war hero. His son Stephen paints a portrait of him as a father. He says, "Dad never quit at anything. He was as mentally tough as he was physically. He despised those that quit. I once found his Lucky Bag yearbook from the Academy and he had written the word 'quit' below the names of classmates who had quit. He struggled to graduate from the Naval Academy. That attitude was instilled in all of us. We were held to high standards and coached never to quit at anything. He was focused on not squandering opportunities. He was a great father... No golf, no going out with friends or other hobbies. When he was home he was totally focused on his family and what we were doing."[29]

In July 1988, Colonel Ripley assumed command of the 2nd Marine Regiment at Camp Lejeune. He then commanded the Navy-Marine Corps ROTC at Virginia Military Institute. During his time at VMI, Col. Ripley created the largest, most productive NROTC unit in the country. He also served as professor of naval science there from August 1990 to June 1992.[30] He retired from the Marine Corps in 1992.[31]

Shortly before his retirement, Ripley appeared before the Presidential Commission on the Assignment of Women in the Armed Forces, testifying against the inclusion of women in the combat arms of the U.S. military.[32] The next year, after his retirement, Ripley appeared before the House Armed Services Committee to testify in favor

Ripley as Colonel

of retaining the traditional ban on service by homosexuals.³³ In both appearances, his testimony was blunt and graphic, based on the realities of combat as he had lived it.

Like many former military men—such as Robert E. Lee, whose picture he'd prayed in front of as a child—Ripley transitioned into academia after retirement. But he seemed to move on every couple of years, almost as if continuing the rotational pattern of his Marine career. In 1996, as president of Southern Virginia College, in Buena Vista, Virginia, he oversaw its transfer to a new board of trustees, primarily of Latter-day Saints.³⁴ (Now Southern Virginia University.)³⁵ In 1997 Hargrave Military Academy in Chatham, Virginia named Ripley as its seventh president.³⁶ In 1999 he became Director of Marine Corps History and Museums and Director of the Marine Corps Historical Center, where he served until 2005.³⁷

If what matters is what we leave behind, John Ripley will stand tall for a long time to come. A bronze statue of him as he appeared at Dong Ha stands at the Marine Corps Museum in Quantico, Virginia. The Class of '67 commissioned an original march in his honor. Composed by Col. Truman W. Crawford, "Ripley at the Bridge" is in the repertoire of the U.S. Marine Drum and Bugle Corps.³⁸ The Ripley Citizen-Leader Scholarship at Southern Virginia University is awarded to the student who exemplifies outstanding citizenship and leadership skills.[39] In 2002, Ripley became the first Marine honored as a Distinguished Graduate of the Naval Academy; that year he also became the only Marine ever inducted into the U.S. Army's Ranger Hall of Fame.[40]

At USNA today, the Ripley Leadership Award is presented to a graduating mid joining the Marine Corps who ranks at the top of his class for physical fitness. In Annapolis, the Ripley Race, a 5K charity event race, is held each November to benefit wounded veterans.[41] His eldest son, Stephen, is Director of Sales for The Program, a team-building and leadership development company founded by Eric Kapitulik '97. His daughter Mary is the Director of Online Content and Media Marketing for the U.S. Naval Institute. His third child, Thomas, works for Blackstreet Capital Management and is one of the co-owners of The Program. Their youngest sibling, John, works for The Marine Corps Scholarship Foundation.

But perhaps more enduring even than these legacies may be words Ripley himself, rather than others, wrote not long before he died. The book is called *A Better Man*, and in it he speaks about what real courage is, and the price one must be willing to pay.[42]

> Physical courage is as close to "tangible' as courage comes. It can be readily seen and readily understood. But there is an altogether different kind of courage—what I would call moral courage—that is rarely seen, even less well understood, but equally deserving of respect and admiration.
>
> Like physical courage, moral courage requires you to make a conscious decision. Though you may not be putting life or limb on the line to act in a morally courageous manner, there are other risks involved that may be very precious. Indeed, like physical courage, moral courage often involves real sacrifice. You must be prepared to give up something that is important to you, and you never really know beforehand the extent of what you may lose.

Ranger Hall of Fame induction, courtesy of Marine Historical Museum

Former Superintendent VADM Richard Naughton '68, USN, escorting Colonel Ripley '66, USMC (Ret.), at the 2002 Distinguished Graduate Award Ceremony

As an exemplar of both kinds of courage, John Ripley fulfilled the highest ideals the Naval Academy strives to inculcate. In that ageless hush of Memorial Hall, may his example, and that incredible moment when he stood alone between victory and defeat, live for mids and grads alike—until the Academy itself is dust, and the memory of heroes fades from the consciousness of what may no longer even call itself human.

14

MEGAN MCCLUNG, ERIK KRISTENSEN, & DOUGLAS ZEMBIEC, CLASS OF 1995: HEROES OF THE WAR ON TERROR

From the first seconds, Naval Academy alumni were among the targets of the Al-Qaeda attack on New York and Washington on September 11, 2001. Several were manning the Navy Command Center and offices elsewhere in the Pentagon; many were killed or grievously injured in the blast, or burned during their escape through flames and falling debris from the upper floors. Other alums—active duty, reservists, retirees, and civilians—died in New York, in the collapse of the Twin Towers, and aboard the airliners commandeered by the terrorists.

A bio of each would be too long for this book, but let us at least remember their names: Charles "Chic" Burlingame III, Class of '71, Kevin P. Connors '69, Gerald F. DeConto '79, Robert E. Dolan, Jr. '81, William H. Donovan, Jr. '86, Patrick S. Dunn '85, Wilson "Buddy" Flagg '61, Kenneth M. McBrayer '74, Michael G. McGinty '81, Jonas M. Panik '97,

Darin H. Pontell '98, Ronald J. Vauk '87, Kenneth E. Waldie '78, and John D. Yamnicky Sr. '52.[1] Their families, friends, shipmates, and classmates will always feel their loss.

From that date on, US forces went to a wartime footing in Afghanistan and follow-on actions throughout the world. In one way or another, probably every grad on active duty since 2001 has been involved. Call it what you will—the Long War, the Global War on Terror (GWOT), Countering Violent Extremism (CVE), even the bland "Overseas Contingency Operations"—it is America's longest conflict, a pitiless, seemingly endless campaign against a hydra-headed threat. The "In Memoriam post-9/11" site maintained by the Alumni Association and Foundation now lists nearly sixty alums killed in operational losses, in aircraft crashes, training accidents, and of course, in action against a determined and resourceful enemy who almost always outnumbers U.S. and Coalition forces.[2]

Each one deserves his or her own article. But we'll focus here on a microcosm that may, in its way, suggest the macrocosm. Three heroes, from the same class. They came of different backgrounds, home towns, religions, service selections, and politics. Contrary to what many think of Academy grads, we're not all precision-molded alike. But these three reflect the experiences of an entire generation of Annapolis men and women in fighting the shape-shifting, deadly enemy of Islamic extremism.

MEGAN McCLUNG, TWENTY-THIRD COMPANY

McClung at Marine Corps Air Station, Cherry Point, NC. Photo courtesy of Re McClung.

Her entry in the USNA archives is brief. Simply, "Megan McClung, Twenty-third Company, General Science Major." Her further record, and those of the other grads in her class, will be closed to the public until the 2060s.[3] But her mother, her classmates, fellow Marines, and friends are more than willing to fill the gap.

At just 5' 4", and weighing only about 125 pounds,[4] Megan Malia Leilani McClung was one of the smallest members of the Class of '95. Her father was an Army brat; her grandfather, a World War II veteran. Michael McClung, Sr. served as a Marine officer in Vietnam during the 1968 Tet offensive, then worked for a defense contractor's classified and unclassified projects[5] before earning a doctorate in information technology.[6] Her mother, Dr. Re McClung, was the daughter of a Navy pilot; McClung herself retired as a school district administrator.[7,8] Megan grew up in Mission Viejo, California. She was on the Cathy Rigby gymnastics team throughout high school, eventually becoming team captain,[9,10] but otherwise stuck close to her homework. Her decision to attend the Academy came as a surprise to her parents. In fact, they didn't find out until she asked them to attend a reception for those who'd just been appointed.[11,12] In her application she had written, "I initially became interested in the Navy at a very early age. I come from a long line of military officers, my father and grandfather were naval officers, and I believe that defending our country is a duty that not only belongs to every man, but to every woman as well."[13] McClung attended a one-year academic prep program at Admiral Farragut Academy in New Jersey, and was their first female graduate[14].

At USNA, she was on the gymnastics team until it was disbanded, after which she joined the diving team.[15] In their Firstie year, Megan and Leah (Lucero) Seay trained together to run the Marine Corps Marathon. Seay says, "I am not a morning person, so the fact that Megan was able to get me out running before classes is nothing short of miraculous. Megan was like that though. She could make a fifteen-mile training run seem fun. In the end, we both completed our very first marathon. I, however, could barely walk down the stairs for a week and Megan was out running again the next day."[16]

Running marathons, "She realized then that being accomplished physically earned her respect that she could leverage as a leader."[17] Megan faltered academically at first, but rallied. Summer school and a change of major enabled her to graduate 888th out of her 917-strong class.[18]

Commissioned as a second lieutenant, USMC, McClung served for nearly ten years at various in-CONUS posts[19], Finding her first choice, flight school, ruled out by chronic airsickness, and barred then by statute from the infantry, she chose combat correspondent/public affairs as promising to get her closest to the action. She later told her dad, "The nicest thing about being a public affairs officer is that I can do everything the infantry guys do, but I don't have to do the paperwork."[20] She served at Camp Pendleton, Parris Island, and Cherry Point.[21]

Megan left active duty in 2004 to join Halliburton (now a subsidiary of KBR), a civilian company mainly carrying out construction and engineering. She worked as a contractor in public affairs in Baghdad, Iraq,[22,23] but

Left to right: Lynn Kinney, Megan McClung, and Amy Forsythe served together during their deployment to Camp Fallujah, Iraq, in 2006. They were assigned to the Camp Pendleton-based I Marine Expeditionary Force (Forward) Public Affairs Office. This photo was taken at Camp Fallujah in April 2006.

remained in the Reserves. When her contract expired, she requested to return to active duty. After a short time at Marine Corps Forces Atlantic, she went back to Iraq as a public affairs officer assigned to I Marine Expeditionary Force Headquarters Group, I MEF, [24], working in Fallujah.[25] When the Army's Ready First Combat Team needed an immediate public affairs office, Megan volunteered. Anxious to lead her own PAO shop, she headed for Ramadi, in dangerous Anbar Province.[26]

Meanwhile, McClung had continued her athletic career, shifting focus from marathons to triathlons. She competed in seven Ironman events, winning the First Military Female award in 2000 at Kona.[27] She organized the first Marine Corps Marathon (Forward) in Fallujah, Iraq to coincide with the 2006 Marine Corps Marathon, and served as the race director.[28] She initiated the Paul the Penguin Award, for the last official finisher at each USMC Marathon,[29] which her family continues to award each year at the race in Washington, DC. Her supervisor in Fallujah said, "Megan could outrun all but four people in camp and she could outshoot everyone not wearing the expert pistol or rifle badge."[30] McClung also continued her academic development, earning a master's in criminal justice from Boston University while stationed in Iraq.[31]

Being a public affairs officer in Anbar, especially in Ramadi—called at the time "The most dangerous city in Iraq,"[32]—was no rear-echelon pogey job. Megan was known for taking care of both her fellow marines and the journalists she shepherded. Once, some enlisted grunts just returned from the field "tired, dirty, hungry—were turned away by the KBR contractor running the mess hall, told 'no food' until they showered. Megan saw that and immediately took KBR to task. Those men got fed. That story about the redheaded captain went rampant, all over, because she understood what the mission was and who was important," her mother says.

One of those journalists, Michael Fumento, gave her this professional tribute: "McClung guided me so I saw what I needed to see rather than what I thought I needed to see. After each embed she diligently provided information that I'd been unable to gather in the field."[33] Another wrote, "She met us at headquarters with a huge grin. 'I get to go out with you guys today! I am so excited!' she gushed. . . 'I'm so glad you're here, so that I can go,' she confided a few minutes later. 'Usually I'm just stuck behind a desk all day. '. . . she couldn't stand to keep still. She went out with her rifle, which looked especially huge on her tiny frame, and a green journal."[34]

McClung was in the final month of her year-long deployment in December 2006.[35] On December 6, she had just dropped off Oliver North (USNA '68) and a Fox News crew after a tour through Ramadi, and was out again escorting a *Newsweek* staffer (in a separate vehicle) on a story about training Iraqi police when a massive roadside bomb exploded. [36,37] Two Army members in her Humvee, Army Capt. Travis Patriquin and Army Spec. Vincent Pomante, III, were also killed.[38]

North wrote ""If everything went as planned, they wouldn't call it 'war.'" That was the tongue-in-cheek assessment of a U.S. Marine Major as to why our helicopter flight from Baghdad to Ramadi had been delayed for half a day. By the time we arrived on the LZ at this outpost of freedom it was the middle of an unusually cold, damp night. A proffered hot cup of coffee was gratefully accepted as the Major helped us load our backpacks, camera gear and satellite broadcast equipment aboard a dust-encrusted Humvee. Just hours later, this widely respected and much admired Marine officer and two brave U.S. Army soldiers were dead, killed by an IED—an improvised explosive device—the insidious weapon of choice for terrorists in Iraq."[39] The Abu Alwan tribe, part of the Awakening, regarded their deaths as an attack on them as well, and killed or captured the insurgents involved in the IED planting within ten days.[40]

Megan's awards include the Bronze Star and Purple Heart.[41] She was buried in Arlington, but lives on in a Marine Corps-wide annual leadership award given in her name, a yearly award by the Sea Service Leadership Association, and the Major Megan M. McClung Memorial Scholarship, awarded by her parents and the Women Marines Association. The Major Megan McClung Memorial Scholarship is sponsored by the Naval Academy Foundation, and there is a scholarship at Boston University in her name as well. The Defense Information School presents the Megan McClung Leadership Award each year, and the Marine Corps Combat Correspondents Association awards the Megan McClung Sport Photography annual merit award to combat correspondents and combat camera specialists.[42]

ERIK KRISTENSEN, SECOND COMPANY

Kristensen wearing his Navy Crew hat after climbing to the top of Mount Katahdin in Maine, 1996. Photo courtesy of Dr. Jonathan L. Bingham

Erik was the son of Edward Kristensen USNA '65, a retired rear admiral. Growing up, he lived in Japan and Guam, among other places.[43] He graduated from Gonzaga College High School, a Jesuit prep school for boys, in Washington D.C. There he "won academic awards but never picked them up," his mother recalled. "He just wasn't a person who wanted the limelight on him." Kristensen played offensive and defensive tackle at Gonzaga. He got the nickname "Spider" because of his spindly legs and arms when down in a four-point stance.[44] He was also trumpet player in the school band, a First Chair, Section Leader, and Co-Concert Master. A "gentle giant with a kind soul and unique sense of humor," Kristensen was an Eagle Scout, too.[45]

At USNA, Spider rowed on the crew team. He raced with the Varsity 4 in three IRA National Championships (93, 94, 95), and raced Freshmen 8 crew at the 1992 IRA National Championships.[46] He earned his N* in heavyweight crew before graduating with an English major and French minor in 1995.[47]

Kristensen applied for SEAL school during his service selection, but was not chosen.[48] After graduation and SWO Division Officer school at Newport, he went to USS *Chandler* (DDG-996) as ordnance control officer, fire control officer, and main propulsion division officer from January 1996 to January 1998.[49] An enlisted shipmate said, "He was

Members of the Class of 1995 who spent all four years rowing Navy Heavyweight Crew (Pete Laschomb, Paul Hockran, Brooks, McFeely, Bill Conner, Mike Lambert, Ed Galloway, Mike O'Hara, Jonathan Bingham, Erik Kirstensen). Photo courtesy of Dr. Jonathan L. Bingham.

among a small handful of officers who were liked and respected command-wide. Personally, as a boatswainsmate, I served with Mr. Kristensen primarily on the bridge and quarterdeck. He made every watch interesting, from quoting the UCMJ to discussing Shakespeare or Melville."[50] His nickname there was "Special K."[51] His next assignment was as Officer in Charge of RHIB Detachment India with Special Boat Team 12 at NAB Coronado, California, from February 1998 to August 1999.[52,53] In 1999, he began teaching English at USNA, while attending graduate school at St. John's College.[54]

But Kristensen had never abandoned his dream of becoming a SEAL. He was finally admitted to BUD/S just under the age limit, as the oldest man in his class. He completed BUD/S Training, Jump School, and SEAL Qualification Training before going to SEAL Team Eight, then as OIC of Alpha Platoon with SEAL Team Ten at NAB Little Creek, Virginia.[55]

In 2005, he deployed to Afghanistan with SEAL Team Ten, becoming the leader of a dedicated team quelling Taliban resistance along the Pakistan border.[56] A friend who served with him during the Red Wings operation said, "Erik leaned to the left. He was liberal in his thinking. Guys gave him a lot of grief but he was witty about it."[57] Other friends recalled him as a "goofball," a "chucklehead," a "down to earth, happy-go-lucky guy,"[58] while also emphasizing his professionalism and dedication. But apparently he was thinking ahead, to a post-SEAL career. Proficient in French, he was to have started a two-year program that fall at the Institute of Political Studies in Paris as an Olmsted Scholar.[59]

But fierce Taliban resistance was to chart a different course for him.

"On June 28, 2005, LCDR Kristensen led a daring mission to rescue a four-man SEAL reconnaissance squad engaged in a fierce firefight with overwhelming Taliban forces in rugged 10,000 foot mountains. Kristensen, seven other SEALs, and and eight Army "Nightstalker" commandos died in the heroic attempt when their MH-47D Chinook helicopter crashed after it was shot down by a rocket-propelled grenade."[60]

The leader of the Taliban force who set the deliberate trap, one "Commander Ismail," was quoted in an interview as saying, "We certainly know that when the American army comes under pressure and they get hit, they will try to help their friends. It is the law of the battlefield."[61]

Kristensen's Bronze Star citation reads, "Demonstrating

exceptional resolve and fully comprehending the ramifications of the mission, Lieutenant Commander Kristensen's element launched aboard a helicopter for direct insertion onto an active battlefield, ready to engage and destroy the enemy in order to protect the lives of their fellow SEALs. While airborne Lieutenant Commander Kristensen continued working with members of his team to develop the plan of attack to support both a Quick Reaction Force and an urgent evacuation of the intended deliberate assault. As the helicopter hovered in preparation for a daring fast-rope insertion of the SEALs, the aircraft was struck by an enemy rocket-propelled grenade fired by Anti-Coalition Militia. The resulting explosion and impact caused the tragic and untimely loss of life of all onboard."

The recon squad fought on, but eventually all but one were killed. The book and film *Lone Survivor* are primarily the story of that single SEAL, Marcus Luttrell, but in the film Kristensen is played by actor Eric Bana. One scene, where he's awakened to go to the rescue of the embattled team, shows Kristensen in his favorite footgear, a touch appreciated by his friends. He was actually buried in his favorite Birkenstocks.[62]

After several narrow escapes from Coalition bombing and strikes targeting his terror cell, "Ismail," (Ahmad Shah) was killed by Pakistani police in 2008 while engaged in a kidnapping.[63]

Kristensen held the Bronze Star with Combat "V" for Valor, Purple Heart, Combat Action Ribbon, and Afghanistan Campaign Medal, awarded posthumously.[64] His funeral services were at the Naval Academy Chapel. He's commemorated with an annual golf outing—the Eye Street Classic—which raises money for the Erik S. Kristensen Memorial Scholarship Fund, helping a Gonzaga student whose family serves in the armed forces.[65] He's remembered at USNA with the Erik S. Kristensen '95

Racing Shell. Erik's classmate and teammate, Brooks McFeely, along with the Naval Academy Foundation, raised funds not just to buy a shell in Erik's name, but to establish an endowment to buy replacements, endowing a rowing shell in the Navy Boathouse forever.[66] The Erik Kristensen Award is presented by the Naval Academy Foundation each year to the Varsity Heavyweight Crew Member of the graduating class who has excelled in Athletics, Leadership, and Academics. The Foundation also administers the Erik Kristensen Scholarship, which sends viable candidates to a post-graduate year of high school before attending USNA.[67,68] The Travis Manion Foundation also commemorated him during USNA's Summer Seminar and USNA STEM (Science, Technology, Engineering, and Math) Camp for rising eleventh graders.[69] And St. John's has an annual memorial lecture series named after him.[70]

Erik was buried in the USNA cemetery near the road facing the boathouse, so he can look down at Hubbard Hall and watch over our current rowers.[71]

DOUG ZEMBIEC, EIGHTH COMPANY

Major Doug Zembiec '95, the "Lion of Fallujah"

Our last alum in this trio is in some ways the best known outside of purely military circles. Born in Kealakekua, Hawaii, the son of an FBI special agent and a grade school teacher,[72] Douglas Alexander Zembiec attended La Cueva High School in Albuquerque, New Mexico. There, he was a state wrestling champion, competing undefeated his senior year.[73]

At USNA, Zembiec majored in political science, and continued to wrestle. He was a Two-time NCAA All-American Wrestler, a NAAA Senior Award Winner at graduation (winning three varsity letters in one sport and participating in that sport for four years)[74], and won the Ed Peery Wrestling Award in 1995 for demonstrating outstanding leadership, dedication, and competitive spirit in the wrestling team.[75]

Doug always pushed himself and his friends, challenging them to do more than they thought possible. His classmate and fellow wrestler Andre Coleman recalls: "... during one of our pre-practice runs we were running from Lejeune to the blinking light up 450 near the intersection of 301/50. Being one of the bigger guys I was lagging towards the back of the group, but Doug stayed back to push me on. On the way back we were making great pace, as we crossed the old Severn River Bridge, Doug looked at me and said we can beat everyone back if we swim to the yard from here. Before I could even contemplate the implications he was

Left to right: *Zembiec; Chris Sanbar '95; Joe Geary '95; and Derek Nelson*

over the rail and in the water, realizing that I would much rather swim than run another mile I was in the water right behind him. Needless to say it took us much longer to swim to the yard than we planned and we therefore were back in the wrestling room only minutes before we had to strap on wrestling shoes and hit the mat."[76]

Zembiec service-selected for the Marine Corps. He served as a Force Reconnaissance platoon commander, being one of the first to enter Kosovo in 1999 as part of Operation Joint Guardian.[77] He deployed to Iraq in Operation Enduring Freedom before taking command of Company E, 2nd Battalion, 1st Marine Regiment, 1st Marine Division in July 2003. He returned to the special operations community in 2004.[78]

The city of Fallujah had been a thorn in the side of the Coalition for years. Its mainly Sunni population had not taken well to the democratic handover of the country to the Shia-majority Interim Government. The First battle of

Fallujah (Spring 2004), also known as Operation Vigilant Resolve, began with the killing of four U.S. security contractors, mutilated and hung on the Highway 10 bridge in west-central Fallujah.[79] Beginning April 6, Echo Company was among those sent in to root out insurgents, launching a month of intense street fighting.[80]

Bing West, a reporter embedded with his unit, called Zembiec "... a wild man, terrific in a firefight and brimming with enthusiasm."[81] ... "Doug Zembiec had a manic grin, as if he wanted to spring up and grab you in a bear hug, maybe breaking a rib by accident, just for the hell of it. 'I'm never so alive,' he told me, 'as in a firefight. Time slows down for me. I can see it all, sense what they're going to do next.'"[82]

He and his Marines fought hard, taking heavy casualties and inflicting many more.[83]

Zembiec "led from the front, rallying his men and directing fire even after being wounded."[84] Ben Wagner (USNA '02) describes how in his first firefight, he "looked across the line of fire, and Captain Zembiec stared back at me and smiled—a reminder that everything was going to be okay."[85]

Along with the Purple Heart, "Zembiec was also awarded the Bronze Star for valor for rushing into the middle of a machine-gun-raked street to get the attention of an Abrams tank supporting Echo Company ... for whatever reason the radio, or "grunt phone," wasn't working, so Zembiec scaled the tank while bullets ricocheted off its hull. After he knocked on one of the hatches repeatedly, the crew of the tank finally opened up. Zembiec then loaded a magazine of illuminated tracer rounds and began shooting from the top of the tank to mark the building from which his Marines were being shot. The tank swung its turret and without warning fired its massive 120mm gun. The blast threw Zembiec into the air and onto the street below."[86]

For his actions in Fallujah, Zembiec was awarded the Bronze Star with valor device and his first Purple Heart.

In interviews, he often said his men had "fought like lions." The title became his: "The Lion of Fallujah."[87,88]

This first battle ended with a Coalition withdrawal in favor of an Iraqi-run security force.[89,90] But that soon crumbled, leaving the city at the mercy of criminals, warlords, and 'Takfiris'—foreign, largely Al-Qaeda-linked radical Islamicists—who earned Fallujah the title "the bomb factory."[91] It became "a safe haven for foreign fighters, terrorists, and insurgents, "a 'cancer' on the rest of al-Anbar Province."[92]

The Second Battle (Operation Phantom Fury) took place in November/December, when the U.S. Army, Marines, Iraqis, and British were ordered in. Their opponents were two to three thousand local insurgents and foreign fighters, stiffened by Chechens who had fought the Russians in Grozny.[93] Many were foreign extremists who were more than willing to be martyred.[94] The six-month withdrawal had permitted them to train, recruit, dig trenches, build defensive berms and roadblocks, prepare VBIEDs, and emplace mines and daisy chains of IEDs.[95] On 8 November, the attack began. Marines cleared buildings room by room behind heavy Army armor.[96] It was called "some of the heaviest urban combat U.S. Marines have been involved in since the Battle of Hue City in Vietnam in 1968."[97] Over the entire Fallujah campaign in 2004, one hundred fifty-one Americans died.[98]

After turning over command, Zembiec was assigned as assistant operations officer at the First Special Operations Training Group, ostensibly conducting predeployment training for deploying Marines. He left there for duty at Headquarters, Marine Corps in mid-2005 and was promoted to major that July. That same year, he married and started a family, wife Pamela and daughter Fallyn.

But the official records don't tell the whole story. Actually, Zembiec never did any staff work at HQMC, nor did he hold a staff position there.[99] That was a cover for his real assignment, to a highly competitive position in

Zembiec, Pamela, and daugher Fallyn

the Ground Branch of the CIA's Special Activities Division. "He went for this with all of his guts and glory," his wife said. "I've never seen this man stressed in my life until he started interviewing for this. He was pacing, and he couldn't sleep."[100] Accepted, he deployed to Afghanistan, then volunteered again to return to Iraq, his fourth deployment to that country, this time to fight not beside Marines, but with Iraqi forces.

On May 11, 2007, Zembiec was leading his unit on a raid in Baghdad. "Family members and former intelligence officials say Zembiec was working with a small team of Iraqis on a "snatch and grab" operation targeting insurgents for capture. Just moments after warning his men that an ambush was imminent, he was shot in the head by an enemy insurgent; he died instantly . . . In the ensuing gun battle, the Iraqis serving beside Zembiec radioed back, "Five wounded, one martyred," according to battle reports."[101] His posthumous citation for the Silver Star read,

"Attacking from concealed and fortified positions, an enemy force engaged Major Zembiec's assault team, firing crew-served automatic weapons and various small arms. He boldly moved forward and immediately directed the bulk of his assault team to take cover. Under withering enemy fire, Major Zembiec remained in an exposed, but tactically critical, position in order to provide leadership and direct effective suppressive fire on the enemy combatant positions with his assault team's machine gun. In doing so, he received the brunt of the enemy's fire, was struck and succumbed to his wounds."[102]

Perhaps the hardest person for any officer to impress is his or her chief or sergeant. In Zembiec's case, his company first sergeant in Fallujah, Bill Skiles, was won over. Skiles said, "He could not handle seeing his marines bleeding and hurting . . . He and I would weep behind closed doors during some of the trying times with mass casualties. Doug's emotions were always worn on his sleeve and I really admired that . . . I cannot name another commander that ALL of his troops would give THEIR lives for if needed. He wasn't fake, he wasn't the most politically correct officer but in the troops eyes that walked the streets with him and fought and sacrificed with him understood.

"Doug's marines loved to laugh with him, cry with him and mostly fight and kill the enemy with him . . . and every marine knew that when Doug shows up to a fight, it makes them feel a little better. Doug allowed the chaplain to perform services during firefights, comforting our grieving warriors after loss and listened to our corpsman on how to take better care of the fallen . . . From his firm handshake to a grieving hug together, I will miss him until I join him. I will even miss the hairiest man on earth, from the eyebrows on down . . . Poor guy had no hair above his eyebrows but he was a human woolly pulley every where

Douglas Zembiec's funeral, Arlington National Cemetery

else. He would try to shave his back before patrols and always miss various spots and yes, I would help finish the job...What are buddies for??"[103]

Retired Marine Corps Colonel John W. Ripley (USNA '62), who was one of those calling on Pamela Zembiec to tell her of Doug's death,[104] called him "absolutely magnetic." "He was a great inspiration, an absolute role model for every one of the Marines he served with," said Col. Ripley, a hero himself who blew up the bridge at Dong Ha in Vietnam. "He would walk into a unit and literally stun every Marine. They would look at him and say, 'My goodness, we got this guy?'"[105]

Zembiec is commemorated by a star chiseled into the Memorial Wall at the CIA headquarters in Langley, and inscribed in the CIA's Book of Honor.[106] The NCAA awarded him the 2008 Award of Valor. The swimming pool located at the Marine Corps' Henderson Hall is named in his honor, and the Commandant of the Corps presents the Douglas A. Zembiec Award for Outstanding Leadership in Special Operations.[107]

In 2008, the Class of '95 conducted the first Run To Honor, a subevent within the Marine Corps Marathon, originally to honor not only the three alums mentioned above, but also three other '95ers killed in the line of duty: Lieutenant Brendan Duffy, USN; Lieutenant Bruce Donald, USN; and Lieutenant Rich Pugh, USN.[108] Run to Honor has since broadened its mission to commemorate, with appropriate athletic events, all USNA graduates who died in the line of duty.

Join or contribute at www.runtohonor.com.[109]

AFTERWORD:
WHAT IS A HERO TODAY?

This was a difficult book to write, especially the last chapter . . . and not just because it's anguishing to contemplate the loss of such promising lives. It returns us to the same questions we faced in the beginning, when we asked what heroes were, and where they came from.

For, in the end, even after examining the biographies of so many brave men and women, the unanswerables remain. Why does one person step up, often again and again, while so many seek the safety of the rear? Why does one military member die, while another in the same squad comes home? What unseen agency selected Howard Gilmore, Megan McClung, Erik Kirstensen, Doug Zembiec, and so many more . . . including the nearly sixty other Academy grads who have died post-9/11?

Perhaps that's a question for the philosophers. But it leads us to another issue. And that is, in a day when drones are controlled from thousands of miles away, and killing takes place from a console, is the nature of heroism itself changing?

I believe it has, along with the technology of weapons, the makeup of our armed services, the face of combat, and the nature of leadership. And it will continue to evolve.

Most of the heroes in this volume achieved fame in battle. But what is the future of battle? In counter-insurgencies, which most of the actions of the war on terror have been, the decisive battle is rare. Our Western distaste for casualties, along with the advance of technology, leads us more and more to decouple personal risk from the battlefield. Command of the air, telepresence, drones, and cyberwar allow us to push violence farther and farther away, both from our operating forces and from the public consciousness.

And on the other side, the "war of evasion and delay," as John Keegan calls it, has become the favorite strategy of our enemies. An uneasy peace is interspersed with sudden, bloody violence, often against soft targets. Its weapons are asymmetrical: the IED, attacks on churches or schools, converting airliners into cruise missiles, turning a presumed ally into an assassin. A reservist in Chattanooga, a public affairs officer in Ramadi, a trooper serving with a psychiatrist in Fort Hood, Texas, a trooper on leave traveling across Europe by train, may suddenly be just as much in battle as Marine infantry in a combat zone.

War will always be dirty, and ugly, and morally compromising. Every conflict in history has taught us this. The noblest crusade has dark chapters. Conflict cannot be sanitized, or rendered bloodless. But, given the nature of these new battlefields, perhaps "heroism" will have to be redefined for the twenty-first century. A hero may not always be a charismatic, virile male warrior who steps forward to volunteer for a dangerous combat mission, rallies troops in battle, receives a grievous wound, gets a medal, and receives an accolade in the newspapers or on CNN.

Or, to put it more clearly, perhaps we must learn that in the times now upon us, heroes may exist outside that traditional algorithm.

Tomorrow's hero may well be any servicemember, selected at what looks at first like random chance. Someone who could just as well be doing a more pleasant, more remunerative job somewhere safer, and spending more time at home with family and friends, spouse and children. But who then, at short notice, or no notice at all, stands by his or her Sailors and Marines, and demonstrates courage and commitment beyond the call of duty.

But perhaps this is not so new. As we mentioned in the introduction, many decorated heroes have said they weren't much different from their shipmates or fellow troops. They just happened to be in the right place at the right time.

When we look at it that way, it seems clear that though some circumstances have changed, the necessity for courage has not. Those who graduate from the Academy today, and those Sailors and Marines they will lead, will be called on to become the next generation of heroes.

To judge by our latest graduates, we don't have to worry. They'll answer that call. Hundreds, even thousands more will step forward to fill the empty boots of those who left us too soon.

There are those who say Americans today are soft, pleasure-loving, and lazy. They say (as our other enemies have before them, about generations before ours) that we've traded our old love of freedom for a culture of safety and an obsession with self. That we've lost the resolution to fight to preserve our way of life. Our freedom to worship, to speak out, and to live as we please. In one word ... our liberty.

How wrong they are.

ENDNOTES

FOREWORD: THE MISSION OF THE NAVAL ACADEMY

[1] Evans, Jule, "Philosophy for Life." <http://philosophyforlife.org/philosophies-for-life/plutarch/>, accessed 17 April 2015.

THE MIDSHIPMEN'S MONUMENT

[1] Cooper, James Fenimore. *Lives of Distinguished American Naval Officers*. Carey and Hart, Philadelphia, 1846. Chapter titled "John Templer Shubrick."

[2] Find a Grave, "Como Irvine Shubrick," < https://www.findagrave.com/memorial/113396274/irvine-shubrick>, accessed 3 Jan 2021.

[3] This is a fictional scene written broadly following the existing records, but with a dose of imagination.

[4] Also see Clark, Paul C. et al, "D-Day Vera Cruz 1847," *Joint Forces Quarterly*, Winter 1995-96. National Defense University, Institute for National Strategic Studies, Washington, DC, 1996.

[5] Bauer, K. Jack, *The Mexican War: 1846-1848*. Macmillan, New York, 1974. Pg. 195.

[6] Find a Grave, "William Branford Shubrick," < https://www.findagrave.com/memorial/37197082/william-branford-shubrick>, accessed 3 Jan 2021.

[7] Sullivan, Naomi, "Monument serves as Testimony to Midshipman Honor," *Trident*, Sept. 14, 2007, pg. 5.

[8] Cheevers, Jim, edited by Sharon Kennedy. *The United States Naval Academy, 1845–2020*. U.S. Naval Academy, Annapolis. <US%20Naval%20Academy%20History%202020%20508.pdf> Accessed 31 Dec 2020.

[9] Benjamin, Park. *The United States Naval Academy, Being the Yarn of the American Midshipman*. Knickerbocker Press, New York, 1900. Pg. 171.

[10] Puleston, W. D. *Annapolis*. Appleton-Century Company, New York, 1942. Pg. 65.

[11] Cheevers, pg. 6

[12] Neufeld, Gabrielle et al, *Marines in the Mexican War*. History and Museums Division, Headquarters, US Marine Corps, 1991, pg. 28.

[13] Bauer, Pg. 106, and others.

[14] Benjamin, pg. 172.

[15] *Servicio de Informacion Agroalimentaria y Pesqueria*. "Un día como hoy, pero de 1847, fuerzas invasoras norteamericanas toman el puerto de Veracruz." Accessed 4 Jan 2021.

[16] Clark, op. cit.

[17] Temple, William. "Memoir of the Landing of the United States Troops at Vera Cruz in 1847." United Service, A Monthly Review of Military and Naval Affairs. L.R. Hamersley, Philadelphia.1896. Vol. XVI, Pg. 421.

[18] Bauer, Pg. 249.

[19] Shock, W.H. "The Castle of San Juan de Ulloa." *The United Service*, April 1897. Pg. 307.

[20] The United Service, Pg. 477.

[21] Bauer, Pg. 249.

[22] Bauer, Pg. 250.

[23] Servicio de Informacion Agroalimentaria y Pesqueria.

[24] The United Service, Pg. 483.

[25] Puleston, Pg. 66.

[26] Temple, Op. Cit.

[27] Servicio de Informacion Agroalimentaria y Pesqueria

[28] Wikipedia, "Revolt of the Polkos." < https://en.wikipedia.org/wiki/Revolt_of_the_Polkos>, accessed 4 Jan 2021.

[29] "U.S. Navy Brig Somers," Naval History and Heritage Command, < https://www.history.navy.mil/browse-by-topic/ships/ships-of-sail/us-navy-brig-somers.html>, accessed 1 Jan 2021.

[30] Cooper, James Fenimore. "Review of the Proceedings of the Naval Court Martial." Henry Langley, New York, 1844.

[31] Wikipedia, "United States Naval Academy." < https://en.wikipedia.org/wiki/United_States_Naval_Academy> accessed 2 Jan 2021.

[32] Somers, op cit.

[33] Find A Grave, "John Ringgold Hynson," < https://www.findagrave.com/memorial/94472869/john-ringgold-hynson>, accessed 3 Jan 2021.

[34] Cheevers, Jim, quoted in Sullivan, Naomi, "Monument serves as testimony to Midshipman Honor," *Trident*, Sept. 14, 2007, pg. 5.

[35] Benjamin, Pg. 172.

[36] Ford, William, unpublished manuscript held by USNA Special Collections and Archives, Chapter 18, page 5. See also Warren, Thomas R., "Operations in California During the Mexican-American War," United States Army School of Advanced Military Studies, United States Army Command and General Staff College Fort Leavenworth, Kansas 2016. However I have been unable to unearth a second source for Ford's third-person account of Mcclanahan.

[37] Bauer, Pg. 121.

[38] Benjamin, Pg. 173.

[39] Benjamin, Ibid.

[40] Naughton, Patrick, "Professional Military Education Proven in Combat during the Mexican War," *Military Review*, Army University Press, July-August 2017.

[41] Leeman, William P. *The Long Road to Annapolis*. UNC Press, 2014. Pg. 226.

[42] Sweetman, Jack. *The U.S. Naval Academy: An Illustrated History*. Naval Institute Press, Annapolis, 1979. Pg. 31. Cited in Fletcher.

[43] Field, Wells L. *One Hundred Years of the U.S. Naval Academy*. Naval Institute Press, Annapolis, 1946. Pg. 36. Cited in research paper by Thomas E. Fletcher, "That Piece of Marble Does Not Look Mexican to Me," April 23, 1986. Held by USNA Archives Division, Nimitz Library.

[44] "Our Oldest Naval Memorial: The Tripoli Monument," Naval Institute Archives, Monday, May 30, 2011. Accessed 1 Jan 2021.

[45] Frazer Design, "Cemetery symbols and the Art of Death," Jul 27, 2016, accessed 3 Jan 2021.

[46] Ford, pg. 11.

[47] Poyer, David. "American Mandarin: The Search for Philo N. McGiffin." *Shipmate*, Sept/Oct 2006, pg. 24.

ROBLEY EVANS

[1] Evans, Robley D. *A Sailor's Log: Recollection of Forty Years of Naval Life*. Smith, Elder & Co., London. 1901. Pg. 14.

[2] Evans, Pg. 18.

[3] Evans, Pg. 21-24.

[4] "Utah War," Wikipedia, <http://en.wikipedia.org/wiki/Utah_War>, accessed 6 Mar 2015.

[5] Evans, Pg. 31.

[6] "Secession at the Naval School." *United States Service Magazine*, April 1864, Pg. 390.

[7] Hayes, John D. "April 1861 - Civil War Comes to the Naval Academy." *Shipmate*, April 1961, pg. 8.

[8] Parker, William Harwar. *Recollections of a Confederate Naval Officer*. Charles Scribner's Sons, New York, 1883. Pg. 161.

[9] Evans, Pg. 39-40.

[10] Falk, Edwin A. *Fighting Bob Evans*. Books for Libraries Press, Freeport, NY. , 1969. (Reprint of original 1931 edition. Page numbers are from the reprinted edition.) Pg. 37.

[11] Falk, Pg. 38.

[12] See Poyer, David, "The Naval Academy's Babylonian Exile," *Shipmate*, March-April 2011, pps. 18-25.

[13] Falk, Pg. 40.

[14] Evans, Pps. 47-50.
[15] Evans, Pg. 54.
[16] Evans, Pg. 60.
[17] Evans, Ibid.
[18] Navsource Online: "USS Powhatan (I)". <http://www.navsource.org/archives/09/86/86036.htm>. Accessed 5 March 2015.
[19] Dillon, Brian Dervin: "The Best-Ever Pistol Shot?" *The Blue Press,* October 2013, pg. 20.
[20] Evans, Pps. 58-62.
[21] "Fighting Bob" Evans at Fort Fisher." Naval History Blog, Naval Institute - Naval History and Heritage Command, posted Saturday, January 15, 2011.
[22] Evans, Pg. 84.
[23] Evans, Pg. 85.
[24] "Fighting Bob," Blog, Op. Cit.
[25] Evans, Pg. 88.
[26] "Fighting Bob" Blog, Op. Cit.
[27] "Fighting Bob," Ibid.
[28] Dillon, pg. 60.
[29] "Fighting Bob" Blog, Op. Cit.
[30] "REAR ADMIRAL ROBLEY D. EVANS, U. S. NAVY, DECEASED," file copy of official USN biographical summary, courtesy Nimitz Library Special Collections, pg. 1.
[31] Evans, Pg. 102.
[32] Evans, Pps. 102-103.
[33] USN bio summary, Op Cit. Quoting Falk.
[34] USN bio summary, Ibid.
[35] Falk, Pg. 95.
[36] Falk, Pps 95-96.
[37] Falk, Pps. 111-128.
[38] Falk, Pg. 129.
[39] Falk, Pg. 130.
[40] Falk, Pg. 137.
[41] Falk, Pg. 139.
[42] USS *Yorktown* CV-10 Association, <https://www.ussyorktown.net>, accessed April 14, 2015.
[43] Evans, Pg. 260.
[44] McSherry, Jack, "Rear Admiral Robley Evans (1846-1912)", the Spanish-American War Centennial Website, <http://www.spanamwar.com/evans.htm>, accessed 15 April 2015.
[45] Falk, Pg. 164.
[46] "Report to Accompany H. R. Bill 11479. House of Representatives Report No. 4079," Feb. 16, 1889. US Govt. Printing Office.
[47] Falk, Pg. 184.

[48] Evans, Pg. 403.
[49] McSherry, Op. Cit.
[50] McSherry, Ibid.
[51] McSherry, Ibid.
[52] *New York Times,* "Pastor Rebukes Capt. Evans," August 11, 1898.
[53] Falk, Pps. 344-345.
[54] Falk, Pg. 410.
[55] National Park Service, "The Roosevelt Pets." <http://www.nps.gov/thrb/learn/historyculture/the-roosevelt-pets.htm> accessed 15 April 2015.
[56] Falk, Pg. 375.
[57] Evans, Robley D. *An Admiral's Log.* D. Appleton & Co., New York, 1910. Pg. 394.
[58] Evans, Admiral's Log, Pg. 392.
[59] Falk, Pg. 424.
[60] Naval History Blog, "The Great White Fleet departs Hampton Roads for Circumnavigation", Dec. 16, 1907. <http://www.navalhistory.org/2012/12/16/december-16-1907-the-great-white-fleet-departs-hampton-roads-for-circumnavigation>, accessed April 16, 2015.
[61] Bailey, Thomas Andrew. *Theodore Roosevelt and the Japanese-American Crises.* Quoting Falk, page 424.
[62] Apparently originally from a letter from Bryce to Grey, Feb. 14, 1908, *British Documents on the Origin of the War.,* VIII, pg. 456; quoted in Falk and Bailey.
[63] Falk, Pg. 425.
[64] "When Chile was kind to 'Fighting Bob," *New York Times,* 7 Jan 1912; courtesy of Nimitz Library Special Collections.
[65] Evans, Admiral's Log. Pps. 433-435.
[66] Evans, Admiral's Log. Pps. 446-7.
[67] Falk, Pg. 437; see also Evans, *Admiral's Log,* Pg.450.
[68] Falk, Pps. 439-442.
[69] USN bio summary. Op. Cit.
[70] Falk, pg. 450.
[71] USN bio summary, Op. Cit.
[72] Personal speculation of this author.
[73] Staton, Michael. *The Fighting Bob.* The Merrriam Press, Bennington VT, 2001. Pg.20.
[74] Text of the PUC.

"SAVEZ" READ

[1] Weems, Bob. *Charles Read: Confederate Buccaneer.* Heritage Books, Jackson, 1982. Pps. 2-4.
[2] Campbell, R. Thomas. *Sea Hawk of the Confederacy.* Burd Street Press, Shippensburg, 2000. Pg. 5.

[3] Campbell, pg. 5.
[4] Jones, Robert A., *Confederate Corsair.* Stackpole Books, Mechanicsburg, 2000. Pps. 19-21.
[5] Campbell, pg. 7.
[6] *Official Register of the Officers and Acting Midshipmen of the United States Naval Academy,* William A. Harris, Washington, 1858. Pg. 9.
[7] *Official Register of the Officers and Acting Midshipmen of the United States Naval Academy,* William A. Harris, Washington, 1859. Pg.8.
[8] Official Register of the Officers and Acting Midshipmen of the United States Naval Academy, George W. Bowman, Washington, 1860. Pg. 7.
[9] "Secession at the Naval School." United States Service Magazine, April 1864, Pg. 390.
[10] Read, Charles W. "Reminiscences of the Confederate States Navy." Papers of the Southern Historical Society. Vol. I.
[11] Read, Ibid.
[12] Campbell, pg. 16-17.
[13] Jones, pg. 30.
[14] Read, Op. Cit.
[15] Jones, pg. 30.
[16] "CSS McRae." <https://www.americancivilwar.com/tcwn/civil_war/Navy_Ships/CSS_McRae.html>, accessed 19 April 2017.
[17] Read, Op. Cit.
[18] Read, Op. Cit.
[19] Jones, pg. 32.
[20] Morgan, James Morris. *Recollections of a Rebel Reefer.* Houghton Mifflin, New York, 1915, Pg. 60.
[21] Pollard, E.A. *Southern History of the War,* Volume I. Richardson, New York,1866. Pg. 299.
[22] Morgan, Pg. 69.
[23] Miller, Edward. *Civil War Sea Battles.* Combined Books, Pennsylvania, 1995. Pg. 30.
[24] Morgan, Pg. 73.
[25] Jones, pg. 173.
[26] Jones, pps. 55-60.
[27] Greene, Francis Vinton. *The Mississippi,* Scribner & Sons, New York, 1882. Pg. 2.
[28] Smith, Mason, *Confederates Downeast.* The Provincial Press, Portland, 1987. Pg. 72.
[29] Silverstone, Paul. *Warships of the Civil War Navies.* Naval Institute Press, Annapolis, 1989. Pg. 202.
[30] "Data Concerning Captain C.W. Read," holograph manuscript held by Mississippi Department of Archives and History, undated, pg. 4

[31] Smith, Myron J. *Civil War Biographies from the Western Waters*. McFarland, 2015. Pg. 199.

[32] Greene, pps. 23-24.

[33] Wikipedia, "CSS Arkansas," < https://en.wikipedia.org/wiki/CSS_Arkansas>, accessed 21 April 2017. Also Read, pg 350-360.

[34] Weems, pg. 72.

[35] Read, pg. 362.

[36] Smith, Myron, pg. 199.

[37] Series I, Volume I, *Official Records of the Union and Confederate Navies in the War of the Rebellion*, Page 768. (Hereafter, *ORN*.)

[38] Miller, Pg. 109.

[39] Campbell, pps. 83-86.

[40] "CSS Florida," Americancivilwar.com, < https://americancivilwar.com/tcwn/civil_war/Navy_Ships/CSS_Florida.html>, accessed 21 April 2017.

[41] Sinclair, Arthur. *Two Years on the Alabama*. Lee and Shepherd, Boston, 1895. Pg. 2.

[42] Campbell, pg. 101.

[43] Silverstone, pg. 218.

[44] Smith, Mason, Pg. 73.

[45] Smith, Mason. Pg. 74.

[46] Tucker, Spencer, et al., *The Civil War Naval Encyclopedia, Vol. I*. ABC-CLIO, 2011. Pg. 687.

[47] Budiansky, *Blackett's War*. Knopf, New York, 2013. Pg. 14.

[48] Jones, pps. 114-166.

[49] Campbell, pg. 112.

[50] Maffit's journal, quoted in Gerard, Philip, "Rebel on the Seas," *Our State*, < https://www.ourstate.com/john-newland-maffitt/>, accessed Aprl 24, 2017.

[51] Jones, pg. 119.

[52] Weems, pg. 108.

[53] Hill, Jim. *Sea Dogs of the Sixties: Farragut and Seven Contemporaries*. University of Minnesota press, Minneapolis, 1935. Pg. 187.

[54] Miller, Pg. 173.

[55] Weems, pg. 113.

[56] Smith, Mason, Pg. 77.

[57] Smith, Mason, Pg. 81.

[58] Campbell, pg. 113.

[59] Hale, Clarence. "The Capture of the Caleb Cushing," paper read before the Maine Historical Society, March 14, 1901. From Nimitz Library Collection.

[60] Shaw, pg. 168.

[61] Shaw, Pg. 168.

[62] Shaw, pg. 170.

[63] Smith, Mason, Pg. 86.

[64] Smith, Mason, Pps. 88-90.

[65] Smith, Mason, Pg.ps. 95-96.
[66] Smith, Mason, Pg. 99.
[67] Shaw, pg. 182-184.
[68] Hale, Pg. 8.
[69] Smith, Mason, Pg. 107.
[70] "Data Concerning," Pg. 7.
[71] Quint, Ryan, "The Battle of Portland Harbor," *Emerging Civil War*, < https://emergingcivilwar.com/2015/02/12/the-battle-of-portland-harbor-conclusion/>, aacessed 21 April 2017.
[72] "Report of Lt. Read, C.S. Navy,"dated July 30, 1863. *ORN*, Series 1, Volume 2, pg. 654-55.
[73] Smith, Mason, Pg. 121.
[74] Shaw, pg. 197.
[75] Shaw, pg. 197-198.
[76] Coski, Pg. 228.
[77] Shaw, Pg. 143.
[78] Hall, James D. "Charles W. Read, Confederate Von Luckner." Unpublished paper held by Nimitz Library, Pg. 9.
[79] Jones, pps. 143-146.
[80] *ORN*, I, 11. Pg. 797.
[81] Coski, pg. 196.
[82] Morgan, Pg. 218.
[83] Coski, pg. 200.
[84] Coski, pg. 209.
[85] Coski, pg. 215.
[86] *Picayune*, New Orleans, "Capt. Charles W. Read," obituary, Jan. 26, 1890. Held at Mississippi Department of Archives and History.
[87] Silverstone, Pg. 231.
[88] Robinson, William M. *The Confederate Privateers*. Yale University Press, 1928. Pps. 244-245.
[89] Weems, pg. 168.
[90] Robinson, Pg. 245.
[91] Jones, pg. 167.
[92] Morgan.
[93] Clark, Charles Edgar, James Morgan, John Marquand. *Prince and Boatswain*. E. A. Hall, New York, 1915. Pg. 69.
[94] Clark, pg. 71.
[95] Clark, pg. 71.
[96] "Data Concerning," Pg. 10.
[97] *Picayune*.
[98] Jones, pg. 173.

FRANK SPRAGUE

[1] Personal email from Professor Samara L. Firebaugh, Chair, Electrical and Computer Engineering Department, United States Naval Academy, 9 Oct 2018.

[2] *New York Herald Tribune*, quoted in Middleton et al. *Frank Julian Sprague: Electrical Engineer and Inventor.* Indiana University Press, 2009. Pg. xv.

[3] Geni. Com, "Frank J. Sprague ,Father of Electric Traction", accessed 2 Oct 2018.

[4] Middleton, William D. at al. *Frank Julian Sprague, Electrical Inventor and Engineer.* Indiana University Press, 2009. Pps 3-4.

[5] Rowsome, Frank Jr. with John L. Sprague. *The Birth of Electric Traction: The Extraordinary Life and Times of Inventor Frank Julian Sprague*, Createspace, 2013. Pg. 19.

[6] Middleton, pg. 7

[7] Geni.com

[8] Personal email from John L. Sprague to author, 23 Oct 2018. Hereafter, "JLS email."

[9] Middleton, pg. 7

[10] Arlington National Cemetery, "Frank Julian Sprague," accessed 3 Oct 2018.

[11] Beach, Edward L. "Frank Julian Sprague," *Army and Navy Life*, May 1908, Pg. 565

[12] Dalzell, Frederick. *Engineering Invention: Frank J. Sprague and the U.S. Electrical Industry.* MIT Press, 2010, Pg. 30.

[13] McBride, William. *Technological Change and the United States Navy, 1865-1845.* JHU Press, 2003.

[14] Rowsome & Sprague, Pg. 21, 76.

[15] Ibid., Pg. 31.

[16] Letter to Sprague from RAdm J. H. Glennon, June 23, 1932. ("Glennon")

[17] Hayes, John D. "Frank Julian Sprague, '78", *Shipmate*, March-April 1962, Pg. 18.

[18] Rowsome & Sprague, Pg. 33.

[19] Dalzell, Pg. 31.

[20] Letter to Sprague from Vadm Harry Huse, 3 June 1932.

[21] Dalzell, Ibid.

[22] Ibid.

[23] Dalzell, pg. 43.

[24] Wikipedia, "USS Richmond (1860)," accessed Oct 3, 2018.

[25] Rowsome & Sprague, Pg. 36.

[26] Glennon, Op. Cit.

[27] Middleton, Pg. 26.

[28] Rowsome & Sprague, Pps. 26, 38.

[29] "Frank Julian Sprague," *Transactions of the American Society of Naval Architects and Engineers*, 1942.

[30] Edison Tech Center, "Generators and Dynamos," and "The Electric Motor," accessed 3 Oct 2018.

[31] Rowsome & Sprague, Pg. 27.

[32] Middleton, Pg. 27.

[33] Ibid., pg. 28.

[34] Commerford, Thomas, "Frank Julian Sprague, Inventor and Engineer," *Scientific American*, Oct 21, 1911.

[35] Rowsome & Sprague, Pg. 32.

[36] Hayes, Op. Cit.

[37] Ibid.

[38] Sprague, Frank J. *Report on the Exhibits at the Crystal Palace Electrical Exhibition, 1882*. Department of the Navy, Bureau of Navigation, Office of Naval Intelligence, Washington, DC, 1884.

[39] Rowsome & Sprague, Pg. 46.

[40] PBS, "The Race Underground." *The American Experience*, 31 Jan 2017.

[41] Rowsome & Sprague, Pg. 44.

[42] Hayes, Op. Cit.

[43] Cummerford, Op. Cit.

[44] Sprague, Harriet. Pg. 13.

[45] Rowsome & Sprague, Pg. 55.

[46] Middleton, Pg. 39.

[47] PBS, Op. Cit.

[48] Cummerford, Op. Cit.

[49] Rowsome & Sprague, Pg. 59.

[50] Ibid., Pg. 40.

[51] National Instruments, "Compound Motors." Mar 30 2016, accessed 3 Oct 2018.

[52] Benjamin, Park. "Sprague Electric Railway & Motor Co., Electric Motor," Modern Mechanism, D. Appleton, New York, 1895, pg.548.

[53] Rowsome & Sprague, Pg.70.

[54] Sprague in Proceedings of the American Electric Railway Association, New York, 1916. Pps. 290-291. Hereafter, "Proceedings."

[55] Rowsome & Sprague, Pg. 70.

[56] Kohlstead, Kurt. "The Big Crapple: NYC Transit Pollution from Horse Manure to Horseless Carriages." 99% Invisible, accessed 3 Oct 2018.

[57] Middleton, Pg. 44.

[58] Middleton, William D. *Metropolitan Railways: Rapid Transit in America*. University of Indiana Press, 2003. Pg. 40. ("Metropolitan Railways")

[59] Sprague, Proceedings. Pg. 292

[60] Middleton, Pg. 70

[61] Rowsome & Sprague, Pg. 79.

[62] Lemelson-MIT, "Frank J. Sprague: Electric Trolley Systems," accessed 3 Oct 2018.
[63] Sprague, Proceedings, Pg. 292.
[64] IEEE Richmond Section, "Milestones:Richmond Union Passenger Railway, 1888," accessed Oct 4, 2018.
[65] PBS
[66] Middleton, Pg. 75
[67] Sprague, Pg. 295.
[68] Rowsome & Sprague, Pg. 89.
[69] Sprague, Pg. 296.
[70] PBS
[71] Hayes, Op. Cit.; Sprague, Pg. 296.
[72] Rowsome & Sprague, Pps.110-116.
[73] Dalzell, Pg. 88
[74] PBS
[75] Rowsome and Sprague, Pps.89-90.
[76] Middleton, pps. 79-83
[77] Edison Tech Center, "Frank Julian Sprague." Accessed 3 Oct 2018.
[78] PBS, Op. Cit.
[79] Lemelson-MIT, Op. Cit.
[80] Rowsome & Sprague, Pg. 5
[81] Dalzell, Pg. 103-112.
[82] Hayes, Op. Cit.
[83] Sprague, Harriet. *Frank J. Sprague and the Edison Myth*. William-Frederick Press, 1947. Pg. 11.
[84] Rowsome & Sprague, Pg. 156.
[85] Middleton, Pg. 87
[86] "Drs. Duncan and Hutchinson appointed Consulting Electrical Engineers to the Rapid Transit Commission." *Electrical Review*. March 16, 1901. Pg.349. See also Middleton, William, "The Frank Sprague Electric Locomotive," *The Railway & Locomotive Historical Society Quarterly*, Winter 2008, Volume 28, Number 1.
[87] Dalzell, pg. 125.
[88] Rowsome & Sprague, Pg. 156.
[89] Philbin, Tom. The 100 Greatest Inventions of All Time. Citadel Press, 2005. Pg. 125.
[90] Sprague, Harriet, Pg. 15.
[91] Jute, Evelyn. "Frank J. Sprague." *Elevator World Unplugged*, July 27, 2016.
[92] Rowsome & Sprague, Pg. 173.
[93] Jute, Ibid.
[94] Dalzell, 142-143
[95] Sprague, Harriet, Pg. 15.
[96] Rowsome & Sprague, Pg. 179.
[97] Jute, Ibid.

[98] Dalzell, Pps. 152-153
[99] "Metropolitan Railways" Pg. 40
[100] Dalzell, pg. 157.
[101] Rowsome & Sprague, Pg. 204.
[102] Dalzell, Pg. 179.
[103] Middleton, pg. 121
[104] Rowsome & Sprague, Pg.219.
[105] Ibid., Pg. 227.
[106] "Metropolitan Railways," Pg. 40.
[107] Dalzell, Pg. 184
[108] Rowsome & Sprague, Pg. 242.
[109] Ibid., Pg. 264.
[110] Sprague, John L. and Joseph J. Cunningham. "A Frank Sprague Triumph: The Electrification of Grand Central Terminal," *IEEE Power and Energy Magazine*, Jan/Feb 2013. Also, Rowsome & Sprague, Pg.247.
[111] Ibid.
[112] Lemelson-MIT, Op. Cit.
[113] Nguyen, Hoa. "Metro-North's 3rd rail was designed for safety." USA Today online, May 8, 2015.
[114] "Daniels Names Naval Advisors," New York Times, Sept. 13, 1915. Page 1.
[115] Rowsome & Sprague, Pg.267.
[116] Dalzell, pg. 225
[117] Scott, Lloyd N. Naval Consulting Board of the United States, US Government Printing Office, 1920. Pg. 201.
[118] JLS email
[119] Camp, Walter Mason. "The Sprague System of Automatic Train Control." *Railway Review*, May 27, 1922. Pps.747-757.
[120] Sprague, Frank. "The Automatic Train Control Problem." *AIEE Journal*, August 1923, Pg.845.
[121] Sprague, John L. Op. Cit.
[122] Dalzell, pg. 229
[123] Sprague, Harriet, Pps. 2-7.
[124] Findagrave.com, "Frank Julian Sprague," accessed Oct 5, 2018.
[125] JLS email
[126] Rowsome & Sprague, Pg. 273.
[127] Massachusetts Museum of Contemporary Art, "History," < https://massmoca.org/about/history/>, accessed Jan 1, 2019.

PHILO MCGIFFIN

[1] "Death of a Brave Man," *The Washington Post*, Feb. 11, 1897.
[2] Forrest, Earle, unpublished manuscript circa 1943? Historical Society of Western Pennsylvania, Heinz Museum, Pittsburgh (main ms. is just "Forrest"

hereafter). Forrest, a feature writer for the *Washington Observer-Reporter*, knew Philo on his return to Washington and interviewed many of his boyhood and Academy acquaintances for the unpublished book.

[3] Richard Harding Davis, *Real Soldiers of Fortune*, Scribner's Sons, New York, 1911. Pg. 122.

[4] Lee McGiffin, *Yankee of the Yalu*, E.P Dutton, New York, 1968. Pg. 15

[5] Beers, J.H. & Co., *Commemorative Biographical Record*, Washington County, Pennsylvania, Chicago, 1893; Pg. 110.

[6] Pittsburgh.about.com, "The Whiskey Rebellion of 1794", accessed May 10, 2006.

[7] References to the elder McGiffin's war service trace back to Forrest, Earle, *History of Washington County Pennsylvania*, 1926, Chicago, S.I. Clarke (in Citizens Library, Washington).

[8] Forrest papers at Heinz Museum, Series 2, Box 19, folder 2.

[9] Lee McGiffin, Pg. 17.

[10] "Descendants of Nathaniel McGiffin," provided by Thomas K. Kaulukukui, Jr.

[11] Forrest, Pg. 64-65.

[12] Lee McGiffin varies, but here I have followed "Descendants of Nathaniel McGiffin."

[13] Lee McGiffin, Pg. 18-19; Forrest Pg. 65.

[14] Forrest, Pg. 66.

[15] Forrest, Pg. 68-69.

[16] "The Great Strike," *Harper's Weekly*, August 11, 1877.

[17] ExplorePAhistory.com, Historical Markers, The Great Railroad Strike of 1877, accessed May 8, 2006.

[18] "Death of a Brave Man."

[19] Forrest, Pg. 71-72.

[20] James Ford Rhodes, "The Railroad Riots of 1877," *Scribner's*, July 1911. Quoted in James W. Latta, *History of the First Regiment Infantry National Guard of Pennsylvania*, J. B. Lippincott, Philadelphia, 1912, pps. 227-228.

[21] *Harper's Weekly*.

[22] Forrest, 72.

[23] Jack Sweetman, *The U.S. Naval Academy: An Illustrated History*, Naval Institute Press, Annapolis, 1979. Pg. 142.

[24] Sweetman, pg. 144.

[25] Letter dated Nov 30, 1943 from Harry L. Hawthorne to Earle Forrest

[26] Letter from Admiral William Fletcher to Earle Forrest, quoted in Forrest's ms.

[27] Richard Harding Davis, *Real Soldiers of Fortune*, Charles Scribner's Sons, New York, 1911. Pg. 123.

[28] Fletcher to Forrest.

[29] Hawthorne to Forrest.

[30] Fletcher to Forrest.

[31] Cyrus Townsend Brady, *Under Tops'l's and Tents*, Charles Scribner's Sons,

New York, 1918. Pg. 50.

[32] Davis, pp 123-124.

[33] Brady, pps. 35-36.

[34] Brady, pg. 37.

[35] Lee McGiffin, 27.

[36] Davis, pg. 124.

[37] Davis, pg. 127.

[38] Earle Forrest, from "records furnished by USNA"; unable to corroborate.

[39] Peter Karsten, *The Naval Aristocracy*, Collier Macmillan, New York, 1972. Pg. 286.

[40] Karsten, pg. 289-90.

[41] Lee McGiffin, 40-45.

[42] Some still to be seen at the Washington County Historical Society.

[43] McGiffin's official US Navy service record.

[44] Lee McGiffin, pg. 37.

[45] Email from Richard Bradford to Poyer, 23 May 2006.

[46] Letter to Poyer from Thomas K. Kaulukukui, Jr., Philo's great grand nephew in Hawaii, 2004.

[47] Philo McGiffin, "China– the Celestial Empire," *Iowa Register* article dated June 10, 1885, pps. 5-7; at Washington Historical Society, Washington, PA.

[48] Richard O. Patterson, "A Commander for China," *Naval Institute Proceedings*; Davis, Pg. 130.

[49] Philo's arrival in Tianjin and interview with Li from his letter to his mother dated April 13, 1885, in Davis.

[50] Patterson, Pg. 1368.

[51] Richard Bradford, "That Prodigal Son: Philo McGiffin and the Chinese Navy," The American Neptune, Vol. 28 (July, 1978), Pg. 160.

[52] Samuel Chu et al, *Li Hung-Chang and China's Early Modernization*, M.E. Sharpe, London, 1994. Pg. 8-9.

[53] Patterson, 1368.

[54] Chu, pg. 21.

[55] Chu, pg. 20.

[56] "China." Encyclopædia Britannica from Encyclopædia Britannica Premium Service. <http://www.britannica.com/eb/article-71777> [Accessed May 10, 2006].

[57] Bradford, pg. 160.

[58] "BookRags Biography on Li Hung-Chang." <http://www.bookrags.com/biography/li-hung-chang/index.html>. Accessed 10 May 2006.

[59] Patterson, 1368.

[60] Philo's letter to his mother.

[61] Email from Richard Bradford to Poyer, 23 May 2006.

[62] Davis, Pg. 132-135.

[63] Richard N.J. Wright, *The Chinese Steam Navy, 1862-1945*. Chatham,

London, 2000. Pg. 27.

[64] Patterson, Pg. 1368.

[65] Patterson, 1369.

[66] Biggerstaff, Knight, *The Earliest Modern Government Schools in China*, Cornell University Press, Ithaca, 1961. Pg. 53.

[67] Some of his students' essays, corrected by Philo, are at the Washington Historical Society.

[68] Lee McGiffin, 104-106.

[69] Lee McGiffin, 85; Samurai History Papers website, "Centers of the Meiji Restoration," http://www.ridgebackpress.com/center.htm. Accessed May 11, 2006.

[70] Some sources state these ships were built by Armstrong, in England. McGiffin may have taken delivery in Britain, but they were actually German, built by the Vulcan yards in Stettin. Wright, Richard N., *The Chinese Steam Navy 1862-1945*, Chatham, London, 2000; also Li Hung-Chang, Memorials, 55:16, cited in Chu, pg. 260; also Armin Wulle, *Der Stettiner Vulcan: Ein Kapitel Deutscher Schiffbaugeschichte*, Herford, 1989, pg. 189. A modern replica of *Ting Yuen* (Dingyuan) is moored in Weihai today.

[71] Wright, pg. 7.

[72] China." Encyclopædia Britannica from Encyclopædia Britannica Premium Service. <http://www.britannica.com/eb/article-71777> [Accessed May 10, 2006].

[73] Wright, Pg. 82.

[74] Chu, Pg. 257.

[75] Lee McGiffin, 107.

[76] Davis, Pg. 139.

[77] Bradford, 163.

[78] Chu, Pg. 256.

[79] Lee McGiffin.

[80] Coat now at the Washington County Historical Society.

[81] Lee McGiffin, 114.

[82] Bradford, 164.

[83] Lee McGiffin, 115.

[84] Wright, Op. Cit. Pg. 89.

[85] Patterson, 1370.

[86] Philo McGiffin, *Century*, pg. 593.

[87] Lee McGiffin, 117.

[88] Wright, Pg. 92.

[89] Bland, J.O.P. *Li Hung Chang*. Henry Holt, New York, 1917. Pg. 232.

[90] John L. Rawlinson, *China's Struggle for Naval Development*, Harvard University Press, Cambridge, 1967. Pg. 166.

[91] Bland, Pg. 223; Spector, Stanley, *Li Hung-Chang and the Huai Army*, University of Washington Press, Seattle, 1964. Pg. 190.

[92] Description primarily from Philo McGiffin, "The Battle of the Yalu," *Century* Illustrated Monthly Magazine, Vol. I - 74, 1895, pgs. 585-604.
[93] Lee McGiffin, 140-141.
[94] P. McGiffin to his mother, letter in Lee McGiffin, Pg. 134.
[95] Laudermilk, Pg. 26.
[96] Laudermilk, 26.
[97] James Allen, *Under the Dragon Flag*. London, William Heinemann, 1898. Pg. 32.
[98] Lee McGiffin, 135.
[99] Rev. J.H. Bausman, "Commander Philo Norton McGiffin, The Hero of the Yalu," *The Navy League Journal*, July 1904, pg. 116.
[100] Bausman, 116.
[101] John Laudermilk, "I Fought at Yalu," *Naval History*, October 1994, Pg. 26.
[102] Patterson, 1374.
[103] "Japan's Great Naval Victory," *New York Times*, Sept 19, 1894.
[104] Rawlinson, Pg. 196, 197.
[105] Laudermilk, Pg. 26.
[106] Biggerstaff, Pg. 58.
[107] Lee McGiffin, 145.
[108] The China Guide: The Summer Palace. <www.beijingbeforetheolympics.com/summer_palace/index.html>, accessed May 10 2006.
[109] Bland, Pg. 220.
[110] Wright, Pg. 95.
[111] Lee McGiffin, 144.
[112] Forrest, 234.
[113] Forrest, pg. 237.
[114] Lee McGiffin, Pg.150.
[115] Bradford, 168.
[116] Forrest, 239.
[117] Forrest, 240.
[118] Nancy Antrim Walker, "The Story of Philo Norton McGiffin," unpublished paper held at the Washington County Historical Society, pg. 14. Walker's father Richard Antrim had corresponded with Robert Thompson, the hospital's director, about Philo in 1929.
[119] Forrest, 240.
[120] Forrest, 240.
[121] Lee McGiffin, 151.
[122] Park Benjamin, "The Story of Philo McGiffin," *Army and Navy Journal*, 20 February 1897.
[123] "Driven to Death," Feb. 11, 1897, article provided by Jim Norton, newspaper name illegible.
[124] "History of NYU School of Medicine," <www.med.nyu.edu/research/res_topics/aboutus/history.html>, accessed May 11, 2006.

[125] "Peter's Field," <www.nycgovparks.org/sub_your_park/historical_signs/hs_historical_sign.php>, accessed May 11, 2006.
[126] Walker, Pg. 15, quoted from letter from Richard M. Thompson to Midn. Richard N. Antrim.
[127] David F. Musto, MD. Schaffer Library of Drug Policy, "The History of Legislative Control over Opium, Cocaine, and their Derivatives." <www.druglibrary.org/SCHAFFER/History/ophs.htm> accessed 11 May 2006.
[128] Lee McGiffin, Pg. 154.
[129] Forrest, from a letter received by Philo's mother after his death, Pg. 245
[130] "Driven to Death" article.
[131] Data on firearms from "Driven to Death" article.
[132] "Death of a Brave Man," Washington Post, Feb. 11, 1897.
[133] "Driven to Death" article.
[134] Forrest, Pg. 247.
[135] Walker, Pg. 15.
[136] Author visit to Washington Cemetery, April 2006.
[137] Forrest, 241.
[138] Forrest, 297.
[139] Norton McGiffin, obituary, *Washington Observer*, Oct 28, 1944.
[140] Benjamin, op. cit.

RICHMOND HOBSON

[1] "Magnolia Grove", flyer by the Alabama Historical Commission, undated.
[2] Thomas McAdory Owen, *History of Alabama and Dictionary of Alabama Biography*, S. J. Clarke Co., Chicago, 1921. III, 821. Cited in Sheldon, page 1.
[3] "The Boyhood of Hobson, Told by his Mother," *Saturday Evening Post*, March 11, 1899. Pg. 585.
[4] H.G. Benners, "The Hero of the *Merrimack*, etc." *Demorest's Family Magazine*, pg. 281.
[5] William Garrott Brown, *The Lower South in American History*, New York, 1903, pg. 233. Cited in Pittman.
[6] Hobson, Richmond P. *Buck Jones at Annapolis*. D. Appleton & Co., New York, 1907. Pg. 3.
[7] Walter E. Pittman, Jr. *Navalist and Progressive: The Life of Richmond P. Hobson*, MA/AH Publishing, Manhattan, KS, 1981. Pg. 8. (Hereafter: Pittman.)
[8] Bennars, op. Cit.
[9] *Congressional Record*, 61st. Congress, 2nd session, 1910, XLV, Pt. 3, 3281.
[10] Harvey Rosenfeld, *Richmond Pearson Hobson: Naval Hero from Magnolia Grove*. Yucca Free Press, Las Cruces, 2001. Pg. 19.
[11] Rosenfeld, pg. 22.
[12] *Congressional Record*, 63rd Congress, 1st session, 1913, L, part 6, 5642.
[13] *Buck Jones*.

[14] Rosenfeld, pg. 23.
[15] *New York Times*, June 7, 1898, 2:4; cited in Sheldon.
[16] RDM Vesey Pratt '89 to Grizelda Hobson, quoted Rosenfeld, pg. 24.
[17] Peter Paret, Gordon Alexander Craig, Felix Gilbert. *Makers of Modern Strategy: From Machiavelli to the Nuclear Age.* Oxford University Press, 1986. Pg. 446.
[18] Pittman, pg. 10.
[19] Generally after Pittman, pg. 10-11.
[20] Richard W. Turk, Introduction to Naval Institute Press edition *of The Sinking of the "Merrimac",* Annapolis, 1987, pg. *xii.*
[21] Turk, pg. *xii.*
[22] Pittman, pg. 11.
[23] Turk, pg. *xiii.*
[24] Turk, *xiv.*
[25] Pittman, pg. 16.
[26] The Sinking of the "Merrimac," pg. 71.
[27] *New York Times*, "Cervera's Fatal Sortie," August 2, 1898, Page 4.
[28] Account of sinking of Merrimac and Hobson's captivity from *The Sinking of the "Merrimac",* The Century Company, New York, 1899; reprinted 1987 by the Naval Institute Press.
[29] Rosenfeld, pg. 65.
[30] Grizelda Hull Hobson, Memoirs. Unpublished manuscript, 1951. Quoted in Rosenfeld, pg. 78.
[31] Rosenfeld, 81.
[32] Rosenfeld, 89.
[33] Quoted in Rosenfeld, 95.
[34] Pittman, pg. 23.
[35] U.S. Naval Station Annapolis website: <http://www.usna.edu/NavalStation/rm_hist.htm>, accessed 19 November 2008.
[36] Richard N. Sheldon, "Richmond Pearson Hobson as a Progressive Reformer," *The Alabama Review,* Oct 1972, Pg. 243.
[37] Pittman, 33.
[38] Sheldon, 244-5.
[39] Rosenfeld, 153.
[40] Sheldon, 245 et seq.
[41] *New York Times*, "Sylvester Shocked at Insults to Women," March 9, 1913, Page 3.
[42] Pittman, 45.
[43] Sheldon, pg. 255.
[44] Hobson, Richmond P., *Alcohol and the Human Race.* Revell, New York, 1919. Pg.182.
[45] Alcohol, Pg. 115.
[46] Hobson, "The Great Destroyer." Speech in Congress, Feb 2, 1911.

[47] Alcohol and the Human Race, pg. 141.
[48] Pittman, 137.
[49] Sheldon, 248.
[50] Walter E. Pittman, Jr. "The Noble Crusade: Richmond P Hobson and the Struggle to Linmit the International Narcotics Trade," *Alabama Historical Quarterly*, Fall/Winter 1972, Pg. 184.
[51] Richmond P. Hobson, "One Million Americans Victims of Drug Habit." *New York Times,* November 9, 1924. Pg. 4.
[52] Pittman, pg. 168.
[53] "The Noble Crusade," Pg. 192-3.
[54] Pittman, 181.
[55] Naval Historical Center, "USS Hobson." <http://www.history.navy.mil/photos/sh-usn/usnsh-h/dd464.htm>. Accessed Nov. 19, 2008.

WENDELL NEVILLE

[1] Richard Collum, *History of the United States Marine Corps*, L. R. Hammersley, New York, 1890, pg. 301.
[2] Alan Millett, *Commandants of the Marine Corps*, Naval Institute Press, 2004. Pg. 214.
[3] Naval Academy Register, 1886-1887. Furnished by Dorothea Abbot of Nimitz Library, USNA, to the author.
[4] Millett, pg. 214.
[5] Naval Academy Register.
[6] Millett, pg. 214.
[7] Charles Lee Lewis, *Famous American Marines.* Page, Boston, 1950. Pg. 233.
[8] Millett, Pg. 216.
[9] Karl Schuon, *The U.S. Marine Corps Biographical Dictionary*. Franklin Watts, New York, 1963. Pg. 159.
[10] Edwin H. Simmons, *The United States Marines: a History.* Naval Institute Press, Annapolis, 2003. Pg. 69.
[11] Millett, Pg. 216.
[12] Simmons, pg.70.
[13] Simmons, Pg. 70.
[14] J. Robert Moskin, *The U.S. Marine Corps Story*, Little, Brown, New York, 1992. Pg. 90.
[15] Millett, Pg.216.
[16] Wilson Graham, "The United States Marine Corps Brevet Medal--One Day Wonder." *Sabretach*, Dec. 1, 1999.
[17] Millett, 216.
[18] Arlington National Cemetery Website, Wendell C. Neville. <www.arlingtoncemetery.net/neville.htm> Accessed Sept 4, 2010.
[19] Andrew Bufalo, *Hard Corps.* S&B Publishing, 2004. Pg. 9. 502 Bufalo, Pg. 8.

[20] George B. Clark, *Treading Softly: U.S. Marines in China, 1819-1949.* Greenwood Publishing Group, 2001. Pg. 40.

[21] Clark, Pg. 41.

[22] Moskin, Pg. 94.

[23] Bufalo, Pg. 8.

[24] Maurice Matloff, *American Military History: 1902-1996.* Da Capo Press, 1996. Pg. 337.

[25] R. G. Price, "Casualties of War - Putting American Casualties in Perspective," November 3, 2003. <rationalrevolution.net/articles/casualties_of_war.htm> accessed Sept 5 2010. Also Hugh Bicheno, "The Philippine Insurrection", in The Oxford Companion to Military History, <www.answers.com/topic/philippine-insurrection>, accessed 5 Sept 2010.

[26] Lillian Powers, *Report of the Twenty-Sixth Annual Conference of the Lake Mohonk Conference*, 1908. Pg. 107.

[27] Sweetman, pg. 65.

[28] Sweetman, pg. 65.

[29] Simmons, Pg. 81.

[30] Millett, Pg. 217.

[31] Moskin, Pps. 201-202.

[32] Robin Cutler, *A Soul on Trial.* Rowman & Littlefield, 2007. Pg. 79.

[33] Arlington National Cemetery Website, James N. Sutton. <www.arlingtoncemetery.net/jnsuttonjr.htm> Accessed Sept 6, 2010.

[34] Mitchell Yockelson, "The United States Armed Forces and the Mexican Punitive Expedition", Prologue (*US National Archives* magazine), Fall 1997, Vol. 29, No. 3.

[35] Burton J. Hedrick, *The Life and Letters of Walter H. Page*, I, 204. Cited by Sweetman.

[36] James W. Hammond, *A Few Marines.* Trafford Publishing, Victoria BC, 2005. Pg. 38.

[37] Tuchman, Barbara, *The Zimmermann Telegram*, pg. 49-50. Cited by Sweetman.

[38] Moskin, Pg. 158.

[39] Sweetman, Jack, *The Landing at Veracruz, 1914.* U.S. Naval Institute Press, Annapolis, 1968. Pg. 67.

[40] London, Jack, "With Funston's Men," *Collier's*, May 23, 1914.

[41] U.S. Army Center for Military History, "Medal of Honor Recipients, Mexican Campaign, Veracruz." <www.history.army.mil/html/moh/mohmex.html> accessed 4 Sept, 2010.

[42] Schuon, Pg. 159.

[43] Lewis, Pg. 236.

[44] Lewis, Pg. 236.

[45] Albertus Catlin, *With the Help of God and a Few Marines.* Doubleday, 1918. Pg. 17+.

46 Lewis, Pg. 237.
47 Gordon Craig, *Europe since 1815*. Holt, Rinehart, & Winston, New York, 1961. Pg. 519.
48 Walter Goerlitz, *History of the German General Staff*. Praeger, 1954. Pg. 193.
49 Robert B. Asprey, *At Belleau Wood*. Putnam's, 1965. Pg. 49.
50 Asprey, Pg. 51.
51 Lewis, Pg. 237.
52 Asprey, Pg. 58.
53 Craig, Pg. 522.
54 Moskin, Pg. 107.
55 Catlin, pg. 83.
56 Lewis, Pg. 238.
57 "Chauchat C.S.R.G. Model 1915 light machine gun." <www.world.guns.ru/machine/mg76-e.htm>, accessed Sept 8, 2010.
58 Lawrence Stallings, *The Doughboys,* Harper & Row, 1963. Pg. 88. (Other sources give other Marines credit for this riposte; but it does sound like Neville's style. Stallings lost a leg at Belleau Wood and later co-wrote the play *What Price Glory*.)
59 McBreen, "2nd Battalion, 5th Marines at Belleau Wood." <www.2ndbn5thmar.com/history/25belleau1918.pdf> accessed 8 sept 2010. Pg. 4.
60 Simmons, Pg. 99.
61 Thomas Boyd, *Through the Wheat*. University of Nebraska Press, 2000.
62 "The Battle for Belleau Wood." <www.worldwar1.com/dbc/ct_bw.htm>. Accessed 9 Sept 2010.
63 Message from Maj. Maurice Shearer to Breg. Gen James Harbord, June 26, 1918.
64 Asprey, pg. 148.
65 "Heaviest losses": from the citation for Valor in Action of Surgeon Paul. T. Dessez, USN, Regimental Surgeon, 5th Marines.; dated July 5, 1918.
66 Floyd Gibbons, *And They Thought We Wouldn't Fight*. Doran, New York, 1918. Cited in Asprey, pg. 168.
67 Asprey, Pg. 279.
68 Lewis, Pg. 249.
69 Moskin, Pg. 123.
70 Millett, Pg. 218.
71 Millett, pg. 218.
72 Schuon, Pg. 159.
73 Moskin, Pg. 188.
74 Millett, Pg. 218.
75 Millett, Pg.223.

MERIAN COOPER

[1] Newsmax, "Foundation Lauds Fighter Pilot, Hollywood Producer Merian C. Cooper." <http://www.newsmax.com/TheWire/Merian-Cooper-Foundation-Illuminate/2012/07/02/id/444241/#ixzz3ee6U3adI>, accessed June 30, 2015.

[2] Jacksonville Historical Society, "Merian Caldwell Cooper: The Man Who Created King Kong," <http://www.jaxhistory.org/portfolio-items/merian-caldwell-cooper-man-created-king-kong/>, accessed Aug 16, 2015.

[3] Biographical Directory of the United States Congress, "Cooper, Charles Merian, (1856-1923)." <http://bioguide.congress.gov>, accessed 30 June 2015.

[4] *King Kong*. RKO Pictures, 1933. DVD from Netflix included audio commentary with snippets of Cooper interviews. Comment on accent by Lenore Hart, native Floridian.

[5] "Jacksonville Historical Society.

[6] *I'm King Kong: The Exploits of Merian C. Cooper*, Turner Entertainment, 2005. Quote from an audio interview with Cooper which is part of this film.

[7] James Thompson. "The Pulaski Legion in the American Revolution." *The Sarmation Review*, April 2005.

[8] Vaz, Mark Cotta, *Living Dangerously: the Adventures of Merian C. Cooper*. Random House, New York. 2005. Pg. 12.

[9] No Coopers are listed in Pulaski's pay records, and the extant accounts of his death do not mention a John Cooper. See Kajencki, Francis, *The Pulaski Legion in the American Revolution*. Southwest Polonia Press, El Paso, 2004.

[10] "Merian Rokenbaugh Cooper, County Judge, Chairman St. Johns Public Schools." Dr. Bronson's St. Augustine History Page, <http://www.drbronsontours.com/bronsoncoopermerianr.html>, accessed June 30, 2015.

[11] The Origins of King Kong, Turner Entertainment, 2005.

[12] *I'm King Kong!*

[13] Vaz, Pg. 11.

[14] *I'm King Kong*. Audio interview with Cooper.

[15] Karolevitz, pg. 19.

[16] Archives Division, Nimitz Library, Cooper file.

[17] "*C*", *Things Men Die For*. Putnam, New York, 1927. Pps. 45-46.

[18] Archives Division.

[19] "C", pps. 22-22.

[20] Limbrick, Peter, "Playing Empire: Settler Masculinities, Adventure, and Merian C. Cooper's The Four Feathers." Monograph, University of California, Santa Cruz, 2009. <http://tlweb.latrobe.edu.au/humanities/screeningthepast/26/the-four-feathers.html#fn9>, accessed Aug 16, 2015.

[21] Vaz, pps. 21-27.

[22] Air Force Historical Research Agency, "20 Bomb Squadron (ACC)." <http://www.afhra.af.mil/factsheets/factsheet.asp?id=9868>, accessed 1 Aug 2015.

[23] Karolevitz, pg. 20.
[24] Warne, Gary C. "Mad Bolsheviks in the US Air Service." <http://warnepieces.blogspot.com/>, accessed 2 Aug 2015.
[25] Lienhard, John H. "The DeHavilland DH-4." University of Houston. < http://www.uh.edu/engines/epi1309.htm>, accessed August 2, 2015.
[26] Rickenbacker, Edward. *Fighting the Flying Circus*. Stokes, New York, 1919, Pps. 321-328.
[27] Goldner, Pg. 24.
[28] Warne, Gary C. "Mad Bolsheviks."
[29] Brigham Young University, "The Movie Man: Merian C. Cooper Before His Career in Cinema," <http://exhibits.lib.byu.edu/the-great-war/the-movie-man.php>, accessed Aug 16, 2015.
[30] Barth, Clarence. *History of the Twentieth Aero Squadron: First Day Bombardment Group, First Pursuit Wing, Air Service, First Army, American Expeditionary Forces.* Battery Press, New York, 1990. (Reprint.) Pg. 38.
[31] Vaz, pg. 39-41.
[32] Hoover Institute, "Memoirs of King Kong Director and War Hero at Hoover," Tuesday, March 4, 2014. <hoover.org/news/memoirs-king-kong-director-and-war-hero-hoover>, accessed 2 August 2015.
[33] Lenin quote from Volkogonov, Dmitri, *Lenin: A New Biography*. The Free Press, New York, 1994. Pg. 329.
[34] Goldner, Orville, et. al; *The Making of King Kong*. A.S. Barnes, Cranbury, NJ. 1975. Pg. 23.
[35] Cisek, Pps. 8-11.
[36] Karolevitz, Pg. 17.
[37] Cooper, letter to Senator Duncan Fletcher, 1920; quoted in Cisek, pg. 45.
[38] "Grass: A Nation's Battle For Life," Press Kit, Milestone Films, 1992.
[39] "Ernest B. Schoedsack," Encyclopedia Britannica, <http://www.britannica.com/biography/Ernest-B-Schoedsack>, accessed 12 August 2015.
[40] Karolevitz, pg. 14-16.
[41] Cisek, pg. 12.
[42] Karolevitz, pg. 21.
[43] Cisek, pg. 44, 52.
[44] Fauntleroy, letter to General Sosnkowski, quoted in Cisek, pg. 61.
[45] Karolevitz, pag. 42-45.
[46] Smithsonian National Air and Space Museum, "Albatros D.Va," <http://airandspace.si.edu/collections/artifact.cfm?object=nasm_A19500092000>, accessed August 12, 2015.
[47] Van Aken, Scott, review of Encore model, <http://modelingmadness.com/scott/w1/encore/72103.htm>, accessed 31 July 2015.
[48] For a good precis, see HistoryNet, "Polish-Soviet War." <http://www.historynet.com/polish-soviet-war-battle-of-warsaw.htm>
[49] Babel, Isaac. *The Complete Works of Isaac Babel.* W.W. Norton, New York,

2002. Pg. 296.

[50] Vaz, pg. 53.

[51] Vaz, pg. 55.

[52] Vaz, pg. 55-57.

[53] Vaz, pg. 60.

[54] Olson, Lynne, and Cloud, Stanley. *A Question of Honor: The Kosciuszko Squadron: Forgotten Heroes of World War II.* Knopf, New York, 2007. Pg. 33.

[55] Cisek, pg. 167.

[56] HistoryNet, "Polish-Soviet War."

[57] Wikipedia, "Marguerite Harrison," <https://en.wikipedia.org/wiki/Marguerite_Harrison>, accessed 12 Aug 2015.

[58] Olson and Cloud, Pps. 32-33.

[59] Cisek, pg. 168-169.

[60] Vaz, pg. 69.

[61] Letter to Capt. Marek Mazynski, quoted in Cisek, pg. 168.

[62] HistoryNet, "Polish-Soviet War."

[63] Vaz, pps. 73-74; p. 324.

[64] Cisek, pg. 201-202.

[65] Solomons Islands Historical Encyclopedia, "Films." <http://www.solomonencyclopaedia.net/biogs/E000108b.htm>, accessed Aug 16, 2015.

[66] Edward A. Salisbury and Merian C. Cooper, *The Sea Gypsy*. G. P. Putnam's Sons, New York, 1924. Pg. 239.

[67] *The Boy Scout's Book of True Adventure*, G.P. Putnam, New York, 1931. Pg. 169.

[68] The Origins *of King Kong*.

[69] Harrison played herself, as a reporter. Wikipedia, <https://en.wikipedia.org/wiki/Marguerite_Harrison>, accessed Aug 13, 2015.

[70] Merian C. Cooper Papers. Biographical History. Brigham Young University. <http://findingaid.lib.byu.edu>, accessed Aug 12, 2015.

[71] Goldner, Pg. 31.

[72] The Boy Scout's Book, pg. 179.

[73] Goldner, Pg. 33.

[74] Goldner, Pg. 34.

[75] Goldner, Pg. 35.

[76] Hambleton, himself a heroic airman of WWI and a friend of Cooper's, died in an untimely death in a plane crash . . . one reason Cooper withdrew from civil aviation. (Note contributed by Mark Cotta Vaz in a personal communication with author, Sept. 23, 2015.)

[77] Merian C. Cooper Papers.

[78] Vaz, pg. 186-187.

[79] Wikipedia, "Edgar Wallace," <https://en.wikipedia.org/wiki/Edgar_Wallace>, accessed 17 Aug 2015.

[80] Goldner, Pg. 31.

[81] The Origins of *King Kong*.

[82] Goldner, Pg. 55-56. Also The Origins of *King Kong*.

[83] Wikipedia, "Merian C. Cooper," <https://en.wikipedia.org/wiki/Merian_C._Cooper>, accessed 12 Aug 2015.

[84] The Origins of *King Kong*.

[85] Goldner, Pg. 167.

[86] Filmography from Wikipedia and Merian C. Cooper Papers.

[87] Merian C. Cooper Papers.

[88] Vaz, pg.263, 275.

[89] Personal communication to author by James D'Arc, 16 Sept 2015.

[90] James D'Arc, 16 Sept, 2015.

[91] Hood, Robert. "Will the Eagles Fly?" 2008. <http://roberthood.net/blog/index.php/2008/03/26/will-the-eagles-fly/>, accessed Aug 16, 2015.

[92] Merian C. Cooper Papers

[93] Chennault, Claire. *Way of a Fighter*. G. P. Putnam's Sons, New York. 1949. Pg. 168.

[94] Chennault, Pps.181-182.

[95] Chennault, Pps. 199-200.

[96] "Cooper '15," *Shipmate*, March 1943, Pg. 62.

[97] "A Clash of Eagles: Stilwell vs. Chennault," Silicon Hutong Archive, <http://siliconhutong.typepad.com/silicon_hutong/2010/01/history-friday-a-clash-of-eagles-stilwell-vs-chennault.html>, accessed 12 Aug 2015.

[98] Tuchman, Barbara. *Stilwell and the American Experience in China*. Macmillan, New York, 1970. Pps. 337 and elsewhere.

[99] Chennault, Pps. 211-212.

[100] Chennault, Pg. 212. But see also Tuchman, Barbara, pg. 338 for a wider discussion of this letter and both sides of the Stilwell-Chennault strategic disagreement.

[101] Vaz, pg. 307.

[102] Vaz, Pg. 313.

[103] *Shipmate*, Pg. 29.

[104] Cooper Papers.

[105] Vaz, pg. 315.

[106] Merian C. Cooper papers.

[107] Personal communication from James D'Arc, 16 Sept 2015.

[108] Vaz, pg.323.

[109] Goldner, pg. 209.

[110] Merian C. Cooper papers.

[111] *The New York Times*, "Dorothy Jordan, 82; Entered Movies in '29", Dec. 13, 1988.

[112] Air Force Historical Society, "412th Test Wing," < http://www.afhra.af.mil/factsheets/factsheet.asp?id=9992>, accessed Aug 13, 2015.

[113] Order of Daedalians,

[114] *Shipmate*, pg. 29.

HOWARD GILMORE

[1] Phone interview by author with William Hagendorn on Oct 26 2009.

[2] Navy Department press release dated May 7, 1943.

[3] Personal communication to the author from Anne Knight, Local History Librarian, Selma-Dallas County Public Library. October 06, 2009.

[4] Gilmore's sworn application, held in USNA Archives.

[5] *Lucky Bag*, 1926, page 520.

[6] From the 1919 *Kaldron*, furnished by Ed Shields and Keven Chatham of Meridian and the Lauderdale County Department of Archives and History.

[7] Phone interview by author with Miss Margaret Scofelia of the Ball H.S. Library.

[8] *Congressional Record*, May 10, 1943; from USN press release May 7.

[9] "Examination for Admission to the United States Naval Academy," exam #5554, Declaration Sheet. Held at Nimitz Library, USNA Archives. Also, page from the USNA entry register filled out by Gilmore, also at USNA Archives.

[10] Relative Standing and Marks Records for Midshipman H.W. Gilmore, USNA Archives.

[11] 1926 *Lucky Bag*, page 520.

[12] Wikipedia, <http://en.wikipedia.org/wiki/USS_Mississippi_(BB-41)>, accessed 19 Jan 2010.

[13] Grider, George. *War Fish*. Little, Brown, Boston, 1958. Pps. 5-6.

[14] Answers.com, biography section. <www.answers.com/topic/hyman-g-rickover>. Accessed Jan 22, 1010.

[15] Fifth Naval District Headquarters Press Release, July 13, 1943.

[16] Blair, Clay. *Silent Victory*. J. B. Lippincott, New York, 1975. Pg. 66.

[17] Whitman, Edward, "Submarine Hero," *Undersea Warfare*, Vol. I No.4, Summer 1999, Pg.22.

[18] Wikipedia, <http://en.wikipedia.org/wiki/USS_S-48_(SS-159)>, accessed 19 Jan 2010.

[19] Friedman, Norman, *Submarine Design and Development*, US Naval Institute Press, 1984. Pg. 39.

[20] Blair, Pps. 64-68.

[21] Friedman, Pg. 39.

[22] Whitman, Pg. 22.

[23] Gugliotta, Bobette. *Pigboat 39*. University Press of Kentucky, Lexington, 1984. Pg. 139; also Grider, Pg. 9.

[24] Blair, above.

[25] Blair, pg. 224.

[26] Grider, Pg. 37.

[27] "History of USS Growler (SS-215)," Navy Department Op-29, Ship's Histories Section.

[28] Gilmore's Navy Cross citation.

[29] "History of Growler."
[30] Gilmore's Gold Star in Lieu of Second Navy Cross citation.
[31] "USS Growler – Report of Fourth War Patrol." March 15 1943.
[32] Fourth War Patrol Report, Pg. 15.
[33] Grider, pg. 38.
[34] Grider, pg. 90.
[35] Howard W. Gilmore Medal of Honor Citation, <www.usna.edu/Admissions/Notables/MOH/bios/gilmore.htm> accessed 21 Jan 2010.
[36] Stan Smith, "Commander Gilmore's Incredible "Suicide" Order," *Stag*, December 1963, Pg. 60.
[37] Blair, Pps. 374-5.
[38] Fourth Patrol Report, pg. 16-17.
[39] "U.S.S. Growler: Report of Casualties." Feb 15, 1943.
[40] W.F. Halsey endorsement on Fourth War Patrol Report, dated Mar 19 1943.
[41] File held at Naval History and Heritage Command, Washington, Naval Warfare Division, incl. King letter dated Mar 18 1943.
[42] New Orleans *Times-Picayune*, May 8 1943.
[43] NY *Herald Tribune*, May 8, 1943. Pg. 26?
[44] Navsea Naval Vessel Register. <www.nvr.navy.mil/nvrships/details/AS16.htm>. Accessed Jan 22 2010.
[45] Letter from Bill Hagendorn to Ward Calhoun dated April 9, 2008.
[46] Email from Chief Wynia of the First Lieutenant's office via Dorothea Abbott, 25 Jan 2010.
[47] Communication from James Cheevers of USNA Museum on Jan 25, 2010.
[48] "Like a Greek tragedy" and notes on family fates: Ward Calhoun, personal communications with author, 26 Oct 2009 and 22 Jan 2010.
[49] Quotes from Hagendorn from phone conversation with author, Oct 26, 2009. Hagendorn died in November, 2009.

VICTOR KRULAK

[1] Sheehan, Neil. *A Bright Shining Lie*. Random House, New York, 1988. Pg. 293.
[2] Other than the sentence from Sheehan, this para is from Coram, Robert, *Brute: The Life of Victor Krulak, U.S. Marine*. Little Brown, New York, 2010. Pps. 22-23.
[3] Grant, Ulysses. S. General Order No. 11. Library of Congress, <http://www.jewishvirtuallibrary.org/jsource/anti-semitism/grant.html>, accessed 12 Sept 2013.
[4] Wikipedia, "History of Antisemitism in the United States, <http://en.wikipedia.org/wiki/History_of_antisemitism_in_the_United_States> accessed 12 Sept 2013.
[5] Dinnerstein, Leonard. *Antisemitism in America*. Oxford University Press,

1994. Pg. 79.

[6] Dinnerstein, Ibid., Pg. 81

[7] Berg, Manfred, *Popular Justice: A History of Lynching in America*, Pg. 134.

[8] Jenkins, Philip, *Hoods and Shirts: The Extreme Right in Pennsylvania, 1925-1950*, University of North Carolina Press, Chapel Hill, 1997. Pg. 114.

[9] Chalmers, David. *Hooded Americanism: The First Century of the Ku Klux Klan*. Doubleday, New York, 1965. Pg. 79.

[10] Personal communication to author, 29 August, 2013.

[11] Coram, Op. Cit., Pg. 30.

[12] "Jewish Ties to the U.S. Naval Academy," Jewish and Israeli News Service, <http://www.jns.org/latest-articles/2012/6/11/jewish-ties-to-the-us-naval-academy.html>, accessed 19 Sept 2013.

[13] Coram, Op. Cit., Pg. 31.

[14] Doyel, Ginger M. *Another Annapolis Alphabet*. US Naval Academy Alumni Association and Foundation. <http://www.usna.com/NC/History/annapolisalphabet/annapolisalphabet4.htm>, accessed 12 Sept 2013.

[15] Personal communication with Victor H. Krulak, Jr., 20 Sept 2013.

[16] "Jewish Ties," Op. Cit.

[17] Dinnerstein, Op. Cit., pg. 75

[18] "Jewish Ties," Op. Cit.

[19] Communication from Dr. Jennifer Bryan, USNA Archives and Records, Nimitz Library, 23 Sept 2013.

[20] Commandant of Midshipmen, "Report of Delinquency in the Case of Midshipman V.H. Krulak, First Class," dated 9 November 1933, courtesy Nimitz Library.

[21] Annual Register of the United States Naval Academy, Annapolis, Md., 1933-34.

[22] Bryan, Op. Cit.

[23] Freeman, Paul, " Abandoned & Little-Known Airfields: Maryland: Anne Arundel County." <http://www.airfields-freeman.com/MD/Airfields_MD_AnneArundelCo.html>, Accessed 23 Sept 2013.

[24] Coram, Op. Cit., pg. 42.

[25] Coram, Ibid., pg. 39.

[26] Annual register of the United States Naval Academy. Annapolis, Md., 1934-1935.

[27] Vandegrift, A. A. *Once A Marine, The Memoirs of General A. A. Vandegrift*, Norton, New York, 1964, pg. 36.

[28] Coram, Op. Cit., pg. 45-46.

[29] Nimitz Library Digital Archives.

[30] Personal letter to author from Rear Admiral Paul Schultz '38 (USN Ret) dated 12 December 2013.

[31] US Marine Corps History Division. "Lieutenant General Victor H. Krulak, USMC (Deceased)." <https://www.mcu.usmc.mil/historydivision/Pages/Who's%20Who/J-L/Krulak_VH.aspx>, accessed Sept 17, 2013.

[32] Krulak, Victor. *First to Fight*. Naval Institute Press, Annapolis. 1984. Pg. 90.

[33] Krulak, Ibid. Pg. 91.
[34] Coram, Op. Cit. pg. 71.
[35] Smith, Holland M. *Coral and Sand.* Zenger, Washington DC, 1948. Pg.47.
[36] Sheehan, Op. Cit. Pg. 295.
[37] Coram, Op. Cit., pgs. 74, 75, 82.
[38] Krulak, Op. Cit, pg. 92.
[39] Smith, Holland, Op. Cit, pg. 90.
[40] Coram, Op. Cit, pg. 95.
[41] Smith, Holland. Op. Cit. Pg. 91.
[42] Adcock, Al, *WWII US Landing Craft in Action.* Squadron/Signal Publications, Carrollton, TX 2003. Introduction.
[43] "Higgins Boats at a Glance." The National WWII Museum in New Orleans. <http://www.nationalww2museum.org/learn/education/for-students/ww2-history/at-a-glance/higgins-boats.html>, accessed 16 Sept 2013.
[44] Roan, Richard W. "Roebling's Amphibian: The Origin Of The Assault Amphibian," USMC Command and Staff College Education Center, Marine Corps Development and Education Command. <http://www.ibiblio.org/hyperwar/USMC/ref/Roebling/Roebling.html> accessed Sept 16 2013.
[45] Roan, Ibid.
[46] Krulak, Op. Cit., pg. 104
[47] Krulak, Ibid., pg. 105.
[48] Christ, James F. *Mission: Raise Hell.* Naval Institute Press, Annapolis, 2006. pgs. 10-11.
[49] Bradsher, Greg, "Operation Blissful." *Prologue* Magazine, Fall 2010, Vol. 42, No. 3. <http://www.archives.gov/publications/prologue/2010/fall/blissful.html> accessed 12 Sept 2013.
[50] Bradsher, Ibid.
[51] Wikipedia, "M1941 Johnson Rifle," <http://en.wikipedia.org/wiki/M1941_Johnson_rifle>, accessed 16 Sept 2013. See also "M1941 Johnson machine gun."
[52] Christ, Op. Cit., Pg. 12.
[53] Lithgow, Shirley. "Seton, Carden Wyndham (1901–1970)." *Australian Dictionary of Biography,* Volume 16, (MUP), 2002.
[54] Christ, Op. Cit., pg. 41.
[55] Christ, Ibid., pg. 86.
[56] Christ, Ibid., pg. 41.
[57] Smith, Holland, Op. Cit. Pg. 16.
[58] Christ, Op. Cit., pg. 52.
[59] Christ, Ibid., pg. 103
[60] Christ, Ibid., pg. 102
[61] Christ, Ibid., pg. 109
[62] Christ, Ibid., pg. 113.

63 Christ, Ibid., pg. 115.

64 Account of the action abstracted from Christ, Ibid.

65 Krulak, Victor, "Preliminary report, Operation BLISSFUL," 5 November 1943, courtesy National Archives.

66 Bradsher, Op. Cit.

67 Bradsher, Ibid.

68 Christ, Op. Cit., pg. 213

69 Moss, William W. "Oral History Interview with LT. GEN. VICTOR H. KRULAK, USMC, November 19, 1970, San Diego, California." Held at the John F. Kennedy Library.

70 *New York Times,* "Victor H. Krulak, Marine Behind U.S. Landing Craft, Dies at 95," January 4, 2009.

71 Coram, Op. Cit., pg. 134.

72 Personal communication to author from Stacey Chandler, Textual Archives, John F. Kennedy Presidential Library, received Sept. 19, 2013.

73 Christ, Op. Cit., pg. 218.

74 Coram, Op. Cit., Pg. 145.

75 Krulak, Op. Cit., pg. 120-121.

76 Coram, Op. Cit., Pg. 161-162.

77 Tucker, Todd. *Atomic America.* Simon & Schuster, New York. Pg. 15. Also cited in *Expeditionary Operations* (Marine Corps Doctrinal Publication 3).

78 Vandegrift, Op. Cit., Pg. 322.

79 "Bombs over Tokyo", <http://www.youtube.com/watch?v=-a-JYmQirQ4> accessed 17 Sept 2013.

80 Coram, Op. Cit. pg. 172.

81 Coram, Ibid. pg. 176.

82 Vandegrift, Op. Cit, pg. 63.

83 Propst, Rodney, Major USMC. "The Marine Helicopter And The Korean War." CSC 1989. <http://www.koreanwaronline.com/arms/VerticalHistory.htm>, accessed 17 Sept 2013.

84 Sheehan, Op. Cit, pg. 295.

85 Krulak, Op. Cit., pg. 122-123.

86 Coram, Op. Cit. Pg. 195.

87 Krulak, Op. Cit. Pgs. 122-124.

88 Krulak, Ibid., pg. 126.

89 Propst, Op. Cit.

90 Propst, Ibid.

91 US Marine Corps History Division, Op. Cit.

92 Reference Section, History and Museums Division, USMC. "Brief History of the Marine Corps during the Korean War." <http://www.kmike.com/MarineCorpsKorea/Marines.htm>, accessed 17 Sept 2013.

93 Propst, Op. Cit.

94 U.S. Marine Corps History Divison, Op. Cit.

95 This paragraph is the result of an email conversation between the author and the three Krulak sons, Vic, Bill, and Chuck, between Sept 25-28, 2013.

96 U.S. Marine Corps History Division, Op. Cit.

97 Sheehan, Op. Cit., pg. 297.

98 Moss, Op. Cit.

99 Coram, Op. Cit., pg. 261.

100 U S Marine Corps History Division, Op. Cit.

101 Halberstam, Op. Cit.

102 Karnow, Stanley. *Vietnam: A History.* Viking, New York, 1983. Pg. 303.

103 "Report of the Office of the Secretary of Defense Vietnam Task Force", Part IV.B.5, "The Overthrow of Ngo Dinh Diem, May-November, 1963," accessed through the National Archives 17 Sept 2013. Declassified per Executi ve Order 13526, Section 3.3, NND Project Number: NND 63316. By: NWD Date: 2011. Pg. iii - iv.

104 Wikipedia, "Krulak Mendenhall Mission," <http://en.wikipedia.org/wiki/Krulak_Mendenhall_mission>, accessed 17 Sept 2013.

105 "Report of the Office of the Secretary of Defense Vietnam Task Force", Op Cit., pg. V.

106 Moss, Op. Cit, pg.7, 19.

107 "Report of the Office of the Secretary of Defense Vietnam Task Force", Op Cit., pg. V.

108 Sheehan, Op. Cit., pg. 305.

109 "Transcript, Roswell Gilpatric Oral History Interview I," 11/2/82, by Ted Gittinger, Internet Copy, LBJ Library. Pg. 6.

110 Krulak, Op. Cit., pgs. 195-198.

111 Krulak, Ibid., pg. 194.

112 Coram, Op. Cit., pgs. 293, 296.

113 Sheehan, Op. Cit. Pg. 630.

114 Krulak, Op. Cit., pgs. 198-199.

115 "President's Daily Diary, August 1, 1966", accessed online from LBJ Presidential Library

116 Sheehan, Op. Cit., pg. 633.

117 Coram, Op. Cit., pg. 313.

118 Krulak, Op. Cit., pg. 202.

119 Halberstam, David. *The Coldest Winter.* Hyperion, New York, 2007. Pg. 478.

120 Lewis Sorley, "To Change a War: General Harold K. Johnson and the PROVN Study," Parameters, US Army War College, Spring 1998, pg. 109.

121 Gilpatric interview, Op. Cit., pg. 15.

122 Coram, Op. Cit., pg. 315.

123 "President's Daily Diary for Jan 27, 1967". Accessed online from LBJ Presidential Library.

124 Sheehan, Op. Cit., pg. 641.

[125] USMC History Division, Op. Cit.

[126] Gates, Robert. Remarks at the Marine Corps Association Annual Dinner, July 18, 2007.

[127] U.S. Naval Academy, "Distinguished Graduates," <http://www.usna.com/page.aspx?pid=365>, accessed Sept 24, 2013.

[128] Henry, Chas, "Marine Corps Lieutenant General Victor H. Krulak, 1913-2008." Naval History Magazine, February 2009, Volume 23, Number 1.

[129] US Marine Corps History Division, "General Charles C. Krulak," <https://www.mcu.usmc.mil/historydivision/Pages/Who%27s%20Who/J-L/Krulak_CC.aspx>, accessed 18 Sept 2013.

PAUL SHULMAN

[1] Personal communication from John Wandres to author, Nov. 10, 2016.

[2] "Interview with Paul Shulman, etc." Oral History by Perry Haber, AACI and American Jewish Committee. May 13, 1993. Held at NY Public Library; access courtesy of AJC. Additional notes by J. Wandres. Throughout, "Oral History"

[3] Ecyclopedia.com, "Dumbarton Oaks Conference." < http://www.encyclopedia.com/social-sciences-and-law/political-science-and-government/united-nations/dumbarton-oaks-conference>, accessed 25 Oct 2016.

[4] Wandres communication, Nov. 10.

[5] Oral History.

[6] Oral History.

[7] Holwitt, J.I. "The Judaic Experience at USNA", USNA Honors Paper, Dept of History, 2003, no. 10. Pps. 18-19.

[8] The title of Shulman's professional notes column in *The Log*.

[9] *Lucky Bag*, Class of 1945, pg. 69.

[10] *Annual Register of the United States Naval Academy, 1943-44*, USGPO, pg. 75.

[11] Communication from J. Wandres to author, 30 Sept 2016.

[12] Oral History.

[13] Wandres, Nov. 10.

[14] DANFS Online, "DD 674", < http://www.hazegray.org/danfs/destroy/dd674txt.htm>, accessed 25 Oct 2016.

[15] Oral History.

[16] DANFS.

[17] Oral History.

[18] Wandres, J. *The Ablest Navigator: Lieutenant Paul N. Shulman, US, Israel's Volunteer Admiral.* Naval Institute Press, Annapolis, 2010, Pg. 23.

[19] "USS Franklin: Struck by a Japanese Dive Bomber During World War II," HistoryNet, <http://www.historynet.com/uss-franklin-struck-by-a-japanese-dive-bomber-during-world-war-ii.htm>, accessed 25 Oct 2016.

[20] Oral History.

[21] Wikipedia, "USS Hunt DD-674", accessed 25 Oct 2016.

[22] Wikipedia, "USS Hunt."
[23] Oral History.
[24] Oral History.
[25] Wikipedia, "USS Hunt."
[26] Wandres, Nov. 10.
[27] Wandres, Nov. 10.
[28] Wandres, Nov. 10.
[29] Greenfield, Murray. *The Jews' Secret Fleet.* Gefen, Jerusalem, 1987. Pps. 35-36.
[30] Wandres, Nov. 10.
[31] "Exodus 1947." The United States Holocaust Memorial Museum. <https://www.ushmm.org/wlc/en/>, accessed Oct 25, 2016.
[32] Communication from Ralph Lowenstein to author, Jan 31, 2017, hereafter, "Lowenstein."
[33] Greenfield, Pg. XV.
[34] Oral History.
[35] Lowenstein.
[36] "Exodus 1947." The United States Holocaust Memorial Museum.
[37] Israel's Navy Special Bulletin, provided by Clandestine Immigration and Naval Museum, Haifa, pg. 6.
[38] Wandres, Pps 45-51.
[39] World Machal, "Paul Shulman."
[40] "Creation of Israel, 1948." Office of the Historian, US State Department. <https://history.state.gov/milestones/1945-1952/creation-israel> accessed Oct 25, 2016.
[41] "1948 Arab–Israeli War," Wikipedia, <https://en.wikipedia.org/wiki/1948_Arab%E2%80%93Israeli_War>, accessed 25 Oct 2016.
[42] Oral History.
[43] Wandres, Nov. 10.
[44] Shulman, Paul, "From My Point of View: Recollections on the Israeli Navy 40 Years Ago," Haifa, December 1987, typescript, Pg. 4 (?).
[45] Aliyah Bet and Machal Virtual Museum, "History of the Navy," <http://www.israelvets.com/hist_navy.html>, accessed Feb. 1, 2017.
[46] Wandres, Pg. 59.
[47] Wandres, Nov. 10.
[48] Lowenstein.
[49] Aliyah Bet and Machal Virtual Museum, "History of the Navy."
[50] Israel's Navy Special Bulletin, pg. 4.
[51] Wandres, Pg. 68.
[52] Navsource Online, "LST-138", <http://www.navsource.org/archives/10/16/160138.htm>, accessed Feb. 1, 2017.
[53] Lowenstein.
[54] Wandres, Nov. 10.
[55] World Machal, "The Sinking of the Egyptian Flagship "EMIR FAROUK."

<http://www.machal.org.il>, accessed 6 Nov 2016.

[56] Wikipedia, "MT Explosive Motorboat," < https://en.wikipedia.org/wiki/MT_explosive_motorboat>, accessed 7 Nov 2016.

[57] Wandres, Nov. 10.

[58] Wikiwand, "MT Explosive Motorboat."<http://www.wikiwand.com/en/MT_explosive_motorboat>, accessed Oct 26, 2015.

[59] Aliyah Bet and Machal Virtual Museum, "History of the Navy."

[60] "Sinking the Emir Farouk," American Jewish Historical Society, < C:\Users\David\Documents\Myfiles\Business (Writing)\Articles - in progress\Paul Shulman\Sinking the Emir Farouk Jewish Virtual Library.htm>, accessed 7 Nov 2016.

[61] World Machal.

[62] "Sinking the Emir Farouk."

[63] World Machal.

[64] Israel's Navy Special Bulletin.

[65] World Machal.

[66] Wandres, Nov. 10.

[67] Greenfield, pg. 32-33, 107.

[68] Shulman, pg. 5.

[69] Wandres, Nov. 10.

[70] Wandres, pps 74-76.

[71] Israel's Navy Special Bulletin.

[72] World Machal.

[73] Wandres, pg. 81-85.

[74] Oral History.

[75] World Machal, "Paul Shulman." < http://www.machal.org.il>, accessed 8 Nov 2016.

[76] Greenfield, Pg. 159.

[77] Oral History.

[78] "Paul Shulman, 72; Headed Israeli Navy," *New York Times* obituary, May 18, 1994.

[79] Wandres, pg. 103.

[80] Tobin, Jonathan. "The Blockade Runners." New Haven Jewish Ledger, Feb. 10, 1995, pg. 20.

[81] Personal communication to author from Hadar Kamon of the Clandestine Immigration and Israeli Navy Museum, Nov. 12, 2016.

[82] Personal communications to author from James Cheevers and David Hoffberger.

[83] Wandres, pg. 103.

THOMAS HUDNER

[1] Hudner, Thomas J. "On a Wing and a Prayer." *Foundation,* Fall 1999, Pg. 28-30.

[2] Tripp, Red, "Hero in Our Midst," *The Valley Patriot,* May 2007. The Lancaster side is from an email from TJH to author dated Feb 23, 2011.

[3] Personal interview of author with Thomas Hudner on 4 Jan 2011. The quoted material below is from Hudner himself unless otherwise noted.

[4] Tripp.

[5] Tripp.

[6] USS Helena website, <http://www.usshelena.org/chew1.html>. Also "U.S. Naval Academy Cemetery and Columbarium" site, <http://www.usna.edu/cemetery/lookup>, both accessed Jan 27, 2011.

[7] "Captain Thomas J. Hudner, Jr." biographic summary from Special Collections and Archives Division, Nimitz Library, USNA. Thanks to Dorothea Abbott.

[8] From Bio summary, Ibid.

[9] Hudner, Thomas. "Jesse's Down." *Foundation* magazine, Spring 1998. Pg. 46.

[10] "USS Leyte, CV-32." From <http://www.navysite.de/cv/cv32.htm>, accessed 25 Jan 2011.

[11] Weems, John E., "Black Wings of Gold." *Naval Institute Proceedings,* July 1983, Pg. 35.

[12] Log of USS *Leyte* for 4 Dec 1950, cited in Weems.

[13] Weems, Pg. 35.

[14] Collins, Michael J. "Mayday over Chosin." *Shipmate,* March 1999. Pg. 10.

[15] Frank, Tim. "Valor in the Frozen Chosin." *Naval Aviation News,* March-April 1998. Pg. 22.

[16] Taylor, Theodore. *The Flight of Jesse Leroy Brown.* Avon, New York, 1998. Summary by author. Also see "Jesse's Down," Pg. 50.

[17] Weems, Pg. 37.

[18] Hudner, "Jesse's Down", pg. 48.

[19] Taylor, Pg. 2; Marine pilot killed, page 4.

[20] "Jesse's Down," pg. 48.

[21] Taylor, Pg. 5.

[22] "Jesse's Down," Pg. 48.

[23] Frank, Pg. 24.

[24] This and following quoted 3 paras from Hudner, "Jesse's Down," Pps. 50-51.

[25] Jesse's Down," Pg. 51.

[26] Clyde McDonnell, "Thomas Jerome Hudner," <http://clydemcdonnell.blogspot.com/2010/08/thomas-jerome-hudner.html>, accessed 30 Jan 2010.

[27] Frank, Pg. 25.

[28] "Jesse's Down," Pg. 52.

[29] Weems, Pg. 38.

[30] Frank, Pg. 25.
[31] Navy Wire Service. "Hudner Takes His Place in History." *Trident*, 7 Feb 1997. From Hudner file at Nimitz Library.
[32] Trident.
[33] Weems, Pg. 38.
[34] Field, James A. Jr. History of U.S. Naval Operations: Korea. Part 9. Naval History Center. <http://www.history.navy.mil/books/field/ch9a.htm> Accessed 30 Jan 2011.
[35] Email communication from TJH to author, Feb 23, 2011.
[36] Harry S Truman Library and Museum, "Desegregation of the Armed Forces: A Chronology," <http://www.trumanlibrary.org/whistlestop/study_collections/desegregation/large/index.php?action=chronology>, accessed 1 Feb 2011.
[37] Email from TJH, Feb 23, 2011.
[38] Truman Library, Record of Appointment Calendar for Apr. 13, 1951. <http://www.trumanlibrary.org/calendar>. Accessed 1 Feb 2011.
859 USNA Public Affairs Office, "Naval Academy Honors Alumni with Distinguished Graduate Award." Story Number: NNS060401-03 Release Date: 4/1/2006.
[39] Email from TJH, Feb 23, 2011.

JOHN RIPLEY

[1] Personal communication from James Cheevers, U.S. Naval Academy Museum, received Aug 16, 2012.
[2] Personal communication from Stephen Ripley to author, Sept. 6, 2012.
[3] Francis Droit Ripley records summary from Special Collections & Archives Division, Nimitz Library, US Naval Academy.
[4] Stephen Ripley, Op. Cit.
[5] NPS, "The Army of Northern Virginia at Gettysburg." <http://www.nps.gov/gett/historyculture/anv-orderofbattle.htm>, accessed 13 Aug 2012.
[6] Stephen Ripley, Op. Cit.
[7] Stephen Ripley, Ibid.
[8] Stephen Ripley, Ibid.
[9] Hevesi, Dennis. "Col. John W. Ripley, Marine Who Halted Vietnamese Attack, Dies at 69." *New York Times*, Nov. 3, 2008.
[10] John W. Ripley Records summary from Special Collections & Archives Division, Nimitz Library, US Naval Academy.
[11] Interview with Minor W. Carter '62, Aug 28, 2012.
[12] Miller, John Grider. *The Bridge at Dong Ha*. Naval Institute Press, Annapolis. 1989. Pg. 13.
[13] JWR records summary from Nimitz Library, Op. Cit.
[14] All quotes from Pete Golwas '62 are from phone interview conducted 25

Aug 2012.

[15] All quoted material from Carter is from interview with author, Aug 28, op. cit.

[16] All quoted material from Madonna is from interview with author, Aug 27, 2012.

[17] Biography of Ripley, John W. Ripley Memorial. <http://www.33usmc.com/Ripley/RipBio.html>. Accessed 14 August 2012.

[18] Wikepedia, "United States Occupation of the Dominican Republic (1965–1966)" <http://en.wikipedia.org/wiki/United_States_occupation_of_the_Dominican_Republic_(1965%E2%80%931966)>, accessed Aug 14, 2012.

[19] Crawford, Danny, et al. *The 2nd Marine Division and Its Regiments.* History and Museums Division, Headquarters, US Marine Corps, Quantico, VA. Pg. 7.

[20] John W. Ripley Memorial, Op. Cit.

[21] "Third Marine Division Patch and History." <http://www.whisperinghope.net/index_files/3rdMarDivHistory.htm>. Accessed 15 Aug 2012.

[22] Garr, Edward. "Third Marine Division Return to Vietnam & I Corps." < http://www.miltours.com/index.php?option=com_content&view=article&id=217:10-23-jun-2012-third-marine-division-return-to-coprs&catid=50:completed-tours&Itemid=62>, accessed 15 August 2012.

[23] Tactical explanation by Arnold Punaro, personal conversation with author. Punaro commanded infantry in the same area later in the war.

[24] Text of official letter of award. Accessed at <http://militarytimes.com/citations-medals-awards/recipient.php?recipientid=4314> on 15 August 2012.

[25] Miller, Op. Cit., pg. 25.

[26] Miller, Op. Cit, Pg. 4.

[27] Ibid. John W. Ripley Memorial.

[28] Miller, Op. Cit.

[29] Stephen Ripley, Sept 6, 2012.

[30] VMI, "Former Professor of Naval Science Dies at 69." Nov. 5, 2008. <http://www.vmi.edu/content.aspx?id=30011> accessed 16 Aug 2012.

[31] Previous three paragraphs summarized from Ripley's official bio.

[32] *The Presidential Commission on the Assignment of Women in the Armed Forces: report to the President, November 15, 1992.* U.S. Government Printing Office, Washington, DC, 1992.

[33] *Policy Implications of Lifting the Ban on Homosexuals in the Military, Hearings before the Committee on Armed Services, U.S. House of Representatives, One Hundred Third Congress.* U.S. Government Printing Office, Washington, DC. 1993. Pps.87-92.

[34] Olsen, Burke. Southern Virginia University, "Former SVC President and Chancellor Passes." November 4, 2008.

[35] Southern Virginia University, 2012 online catalog, <http://svu.edu/about> accessed 16 Aug 2012.

[36] Hargrave Military Academy, "Hargrave History." <http://www.hargrave.

edu/admissions/about-hargrave/history/> accessed 16 Aug 2012.

[37] Wikipedia, "United States Marine Corps History Division." <http://en.wikipedia.org/wiki/United_States_Marine_Corps_History_Division>, accessed 12 Sept 2012.

[38] Thomas E. Moncrief. *Colonel Truman W. Crawford, director/commander of "The Commandant's Own" U.S. Marine Drum and Bugle Corps*. University of South Carolina, 2006. Pg. 26-30. Dissertation, accessed on ProQuest 14 Aug 2012.

[39] SVU catalog, Pg. 11.

[40] John Grider Miller, Eulogy of Ripley, <http://www.usni.org/colonel-ripley>, accessed 12 Sept 2012.

[41] Personal communication from Thomas Ripley on 6 Sept, 2012.

[42] Johnson, Kelly H. *A Better Man: True American Heroes Speak to Young Men on Love, Power, Pride and what It Really Means to be a Man*. Brandylane Publishers, Inc., Richmond, VA 2009. Pg. 49 et al.

HEROES OF '95

[1] "In Memoriam", Naval Acaemy Alumni Foundation, <http://www.usna.com/page.aspx?pid=605>, accessed Jan 26 2016.

[2] "In Memoriam: Operational Losses and KIAs Post-9/11," USNAAA and Foundation, <http://www.usna.com/page.aspx?pid=503>, accessed Jan 28 2016.

[3] Communication to Author from USNA Archives Division, Nimitz Library

[4] Barber, Mike, *Seattle Post-Intelligencer*, via AP and *Military Times*, "Honor the Fallen,' <http://thefallen.militarytimes.com/marine-maj-megan-m-mcclung/2419533>, accessed 26 Jan 2016.

[5] Barber, Op. Cit.

[6] Dr. Re McClung, personal communication to author, Feb. 8, 2016.

[7] Barber, Op. Cit.

[8] McClung, Op. Cit.

[9] Thomas, Joseph J. *Leadership Embodied*. Naval Institute Press, Annapolis, 2013. Chapter 53, "Presence," by Gerg Overbeck and Wesley S. Huey. Pg. 208.

[10] McClung, Op. Cit.

[11] Barber, Op. Cit.

[12] Thomas, Op. Cit. Pg. 208.

[13] Thomas, Op. Cit. Pg 209.

[14] Communication to author from Dr. Re McClung, Jan 9, 2016.

[15] Communication to author from Linda Seay, March 2, 2016.

[16] Thomas, Op. Cit., Pg. 209.

[17] Thomas, Op. Cit., Pg.209.

[18] Story, Courtesy. "Be brief. Be bold. Be gone: A decade later, Maj. Megan M. McClung's legacy lives on," press release, Headquarters Marine Corps, Dec. 22, 2015. <http://www.marines.mil/News/NewsDisplay/tabid/3258/

Article/637723/be-brief-be-bold-be-gone-a-decade-later-maj-megan-m-mcclungs-legacy-lives-on.aspx>, accessed Jan 27, 2016.

[19] Thomas, Op. Cit, Pg. 209.

[20] Thomas, Op. Cit., Pg. 209.

[21] Ritchie, Erika. "O.C. Native Dies in Iraq." *The Orange County Register*, Tuesday, December 12, 2006. < http://www.arlingtoncemetery.net/mmmcclung.htm>, accessed 27 Jan 2016.

[22] McClung, Op. Cit.

[23] "Killed in Iraq or in support of Operation Iraqi Freedom," <http://www.nooniefortin.com/iraq.htm>, accessed 27 Jan 2014.

[24] McClung, Op. Cit.

[25] McClung, Op. Cit.

[26] McClung, Op. Cit.

[27] Story, Op Cit.

[28] "Marine Corps Marathon's Infamous Penguin Award," *USMC Life*, <http://usmclife.com/2015/01/marine-corps-marathons-infamous-penguin-award/>, accessed 27 Jan 2016.

[29] Thomas, Op. Cit., Pg. 210.

[30] McClung, Op. Cit.

[31] Fumento, Michael. "In Memoriam: Farewell to Maj. Megan McClung, USMC," *The American Spectator*, 12.26.06, <http://spectator.org/articles/46023/farewell-maj-megan-mcclung-usmc> accessed 27 Jan 2016.

[32] Fumento, Michael. "Maj. Megan McClung and Capt. Travis Patriquin, RIP," Weblog, <http://www.fumento.com/weblog/archives/2006/12/>, accessed 27 Jan 2016.

[33] Thomas, Op. Cit., Pg. 213.

[34] "Killed in Iraq," Op cit.

[35] Statement by PAO at Camp Fallujah, 18 Dec 2006, <http://www.arlingtoncemetery.net/mmmcclung.htm> accessed 26 Jan 2016.

[36] Thomas, Op. Cit., Pg. 213.

[37] Story, Op. Cit.

[38] North, Oliver, "State of War II", December 8, 2006, <http://townhall.com/columnists/olivernorth/2006/12/08/state_of_war_ii/page/full>, accessed March 14, 2016.

[39] Thomas, Op. Cit., Pg. 213.

[40] Thomas, Op. Cit., Pg. 214.

[41] McClung, Op. Cit.

[42] Story, Op. Cit.

[43] Lengel, Allan, "Navy SEAL From the District Died Leading Rescue Mission," *The Washington Post*, July 6, 2005. < http://www.washingtonpost.com/wp-dyn/content/article/2005/07/06/AR2005070600057.html>, accessed Jan 28, 2016.

[44] Lengel, Op. Cit.

[45] "About Erik S. Kristensen," The Erik S. Kristensen Eye Street Classic, May

6, 2016. <http://kristensenklassic.com/about-erik/>, accessed 28 Jan 2016.

[46] Communication to author from Jennifer Erickson, USNA PAO, dated Jan 29, 2016.

[47] Eye Street Classic, Op. Cit.

[48] Meek, James G., "An Overlooked Hero of Navy SEALs' Operation Red Wings," ABC News, Jul. 1, 2015. <http://abcnews.go.com/International/overlooked-hero-navy-seals-operation-red-wings/story?id=32136944>, accessed Jan 28, 2016.

[49] Veteran Tributes: Honoring Those Who Served. <http://www.veterantributes.org/TributeDetail.php?recordID=1971>, accessed 28 Jan 2016.

[50] Pulliam, Steve, note to Fallen Heroes Memorial, <http://www.fallenheroesmemorial.com/oef/profiles/kristeneriks.html>, accessed 28 Jan 2016.

[51] A QM2, note to Fallen Heroes Memorial.

[52] Veteran Tributes, Op. Cit.

[53] "Erik. S. Kristensen," Navyseals.com. <http://navyseals.com/nsw/erik-s-kristensen/>, accessed 28 Jan 2016.

[54] Obituary, Op. Cit.

[55] Veteran Tributes, Op. Cit.

[56] "Erik. S. Kristensen," Navyseals.com. <http://navyseals.com/nsw/erik-s-kristensen/>, accessed 28 Jan 2016.

[57] Meek, Op. Cit.

[58] Meek, Op. Cit.

[59] The Capital, 6 July 2005

[60] "Erik. S. Kristensen," Navyseals.com. <http://navyseals.com/nsw/erik-s-kristensen/>, accessed 28 Jan 2016.

[61] Myers, Lisa, "An Interview with a Taliban Commander." NBC News, Jan 27, 2005, <http://www.nbcnews.com/id/10619502#.VqpB0sf2bAU>, accessed 28 Jan 2016.

[62] Ismay, John. "Seeing My Friend Depicted in 'Lone Survivor'," the *New York Times*, January 24, 2014. <http://atwar.blogs.nytimes.com/2014/01/24/seeing-my-friend-depicted-in-lone-survivor/?_r=0>, accessed 28 Jan 2016.

[63] Dupee, Matt, "Bara bin Malek Front commander killed in Pakistani shootout," *The Long War Journal,* April 17, 2008. <http://www.longwarjournal.org/archives/2008/04/_commander_ismail_im.php> accessed 28 Jan 2016.

[64] "Eric S. Kristensen," Op Cit.

[65] The Eye Street Classic, Op. Cit.

[66] "Hands On the ERIK KRISTENSEN," *Shipmate*, Jan-Feb 2008, pps. 4-5.

[67] Communication from Robert Friedrich, Director of Rowing, Head Men's Coach, U.S. Naval Academy, dated Jan 29, 2016.

[68] Navy Crew Newsletter, February 2014, pg. 3.

[69] "Fallen Heroes," Travis Manion Foundation, <http://www.travismanion.org/hero/lcdr-erik-s-kristensen-usn/>, accessed Jan 28, 2016.

[70] St. John's College Newsletter, Feburary 2015, <https://www.sjc.edu/

files/9414/2676/8252/Kristensen_Newsletter_2015.pdf>, accessed 28 Jan 2015.

[71] Communication from Robert Friedrich.

[72] Gibbons-Neff, Thomas. "Legendary Marine Maj. Zembiec, the 'Lion of Fallujah,' died in the service of the CIA." *The Washington Post,* July 15, 2014.

[73] Douglas A. Zembiec, Wikipedia, <https://en.wikipedia.org/wiki/Douglas_A._Zembiec>, accessed 31 Jan 2016.

[74] USNA Archives, Nimitz Library.

[75] Communication to author from USNA PAO Jennifer Erickson on Jan 28, 2016.

[76] Communication to author from Andre Coleman on Feb. 16, 2016.

[77] Wikipedia, Op. Cit.

[78] US Marine Corps History Division, biography, USMC Portal, via Wikipedia. <https://en.wikipedia.org/wiki/Portal:United_States_Marine_Corps/biography/2009April>, accessed 29 Jan 2016.

[79] Estes, Kenneth W. *U.S. Marines in Iraq, 2004–2005: Into the Fray,* History Division, United States Marine Corps, Washington,DC. Pg. 30.

[80] Blackfive, "Showdown Part 8," Tuesday, August 03, 2004. < http://www.blackfive.net/main/2004/08/showdown_part_8.html>, accessed 20 Jan 2016.

[81] West, Bing. *The Strongest Tribe.* Random House, New York, 2008.Pg. 91.

[82] *The Strongest Tribe*, Op. Cit., pg. 360.

[83] See West, Bing, *No True Glory*, Bantam, New York, 2005, Pps. 194-207 for excellent description of Zembiec in action on April 24, 2004.

[84] Gibbons-Neff, Thomas. "Legendary Marine Maj. Zembiec, the 'Lion of Fallujah,' died in the service of the CIA." *The Washington Post,* July 15, 2014.

[85] Welles, Joshua, et al, *In the Shadow of Greatness,* Naval Institute Press, Annapolis, 2012. Pps. 98-99.

[86] Gibbons-Neff, Op. Cit.

[87] Siegel, Andrea. "Famed 'Lion of Fallujah' dies." *The Baltimore Sun*, May 14, 2007.

[88] Gibbons-Neff, Op. Cit.

[89] Matthews, Matt M. "Operation AL FAJR: A Study in Army and Marine Corps Joint Operations," Combat Studies Institute Press, Fort Leavenworth, Kansas, 2006. Pps. 9-10.

[90] Estes, Op. Cit. Pg. 37.

[91] West, Op. Cit., pg. 228. "Bomb Factory" quote, pg. 248.

[92] Estes, Op. Cit., Pg. 55.

[93] Strongest Tribe, Op. Cit, pg. 55.

[94] Matthews, Op. Cit., pg.40.

[95] West, pps, 256-257; Matthews, pg. 40.

[96] Mathews, pg. 67.

[97] "2 Battle of Fallujah (Phantom Fury): A Selected Bibliography," Morris Swett Library, February 2014. <http://sill-www.army.mil/MorrisSwett/Fallujah.pdf>, accessed 31 Jan 2016.

[98] The Strongest Tribe, pgs. 59-60.

[99] Ben Wagner, personal communication to the author, Feb. 5, 2016.
[100] Gibbons-Neff, Op. Cit.
[101] Gibbons-Neff, Op. Cit.
[102] Quoted from award citation
[103] "Remembering the Lion of Fallujah - Major Doug Zembiec," Black Five, Wednesday, May 16, 2007. <http://www.blackfive.net/main/2007/05/remembering_the.html>, accessed 29 Jan 2016.
[104] Zembiec, Pamela. *Selfless beyond Service*. Outskirts Press, 2014. Pg. 2.
[105] Arlington National Cemetery Website. <http://arlingtoncemetery.net/dazembiec.htm>, accessed 29 Jan 2016.
[106] Gibbons-Neff, Op. Cit.
[107] Wikipedia, Op. Cit.
[108] Dooley, Katie, "Running in their Honor," Shipmate, Jan-Feb 2008, Pps. 6-8.
[109] Communication to the author from D. K. Richardson, Jan 31, 2016.

ABOUT NORTHAMPTON HOUSE PRESS

Northampton House LLC publishes selected fiction, lifestyle nonfiction, memoir, and poetry. Watch the Northampton House list at www.northampton-house.com, and Like us on Facebook—"Northampton House Press"—to discover more innovative works from brilliant new writers.

www.ingramcontent.com/pod-product-compliance
Lightning Source LLC
Chambersburg PA
CBHW030146100526
44592CB00009B/142